Please Tell Me,

WHY

Can't They Stop!?!

Please Tell Me,

WHY

Can't They Stop!?!

UNDERSTANDING ADDICTION

Randall and Alice R.

authorHOUSE®

AuthorHouse™ LLC
1663 Liberty Drive
Bloomington, IN 47403
www.authorhouse.com
Phone: 1-800-839-8640

Published by AuthorHouse   11/14/2013

ISBN: 978-1-4918-2434-4 (sc)
ISBN: 978-1-4918-2433-7 (e)

Library of Congress Control Number: 2013918044

DEDICATION

This book is dedicated to Roberta Temple. Roberta was a leader in the recovery community, and a brave lady with a huge heart. Roberta recently passed away—but the results that she achieved in life will last forever.

This book is also dedicated to the great people at Russ Recovery and The Betty Ford Center.

Finally, this book is dedicated to the Alcohol and Drug Studies and the Addiction Studies programs at College of the Desert and Diablo Valley College, respectively. These programs train people who save lives and families every day.

CONTENTS

PREFACE

We have graduate degrees from top schools, and we worked at Fortune 500 companies. However, we couldn't use our existing knowledge to address our son's addiction—we had to acquire new knowledge and skills, and develop our ability to apply the new knowledge and skills in real-life situations. The purpose of this book is to share this new knowledge and the skills with you.

Let us begin with the obvious: We cannot solve our son's addiction because addiction is our son's life-long disease. There is no cure and no 100% successful treatment program. If someone is trying to sell you a "cure"—save your money because a "cure" does not exist. Think about it—if a cure for addiction did exist, why isn't everyone using the cure and why are there so many addicts? On the other hand, addiction is a manageable disease. Our son can choose to manage his addiction. We cannot make that choice for him, <u>but we can choose how we deal with our son's addiction</u>. This book is for you—the "normy"—so you can better deal with your family member's or your friend's addiction.

To better ensure that you receive useful and accurate information, we asked Russ Recovery to review this book. Russ Recovery, based in Palm Springs, is an intensive outpatient program that uses an integrated behavioral health model to treat addiction. The program includes medical, psychological, behavioral and 12 step components. Our son participated in many recovery programs, and he learned from each of them. There are many good addiction recovery programs. However, at Russ Recovery, our son developed a more complete toolbox of skills, he learned to apply these skills under the guidance of knowledgeable and caring professionals, he received excellent medical treatment from a skilled addiction physician and team, and he finally learned to make recovery the most important thing in his life. We hope that our son remains sober for his entire life; but, if he does relapse, we know that he now has the skills and experience to regain his sobriety.

Our goal in this book is to introduce you to the illness of addiction, and to guide you to make better choices. We want to condense our experiences and knowledge into a book that you can quickly and readily use—so that you can improve your knowledge and outcomes. Your ability to use and apply this new knowledge may be initially problematic, but your ability should improve as you use the knowledge in real life situations.

To create this short book for you, we relied on some excellent reference books. The reference books are identified in the endnotes. We recommend that you purchase these reference books because they will develop your understanding of addiction. The more informed that you become, the better you will be able to manage the disease of addiction.

PLEASE TELL ME—WHY CAN'T THE ADDICT STOP?

The Answer.

Why can't the addict stop? The simple answer is that (1) drug use changed the addict's brain and (2) the addict's brain produces thoughts that encourage drug use. The physical body is rewarded by the euphoria, feelings and memories experienced through drug use, and the brain learns to value and reward drug use. The changed brain adapts to drug use, it becomes less able to produce pleasure and rewards from non-drug experiences, and it becomes less able to produce normal thoughts or to exercise normal judgment. When you ask, "Why can't you stop?", the addict hears your question; but, the addict may not see the need to stop. The addict's brain produces thoughts that <u>encourage</u> them to continue their drug use, and the brain produces pleasure and pleasurable memories from the drug use. The addict's answer to your question is probably, "There is no need to stop. Why should I stop?"

The addict does not see drug use as the problem—it is their solution. Drug use allowed the addict to experience pleasure, to cope with life's challenges, and to address their anxiousness and pain; and, even though drug use may cause some problems, the addict considers drug use as an effective solution. From the addict's perspective (i.e. <u>the thoughts produced by the addict's brain</u>), drug use—even though it produces

1

some negative problems and consequences—is the *answer* to their problems rather than the source of their problems.

Let me restate the problem here again: The problem is NOT the drug use. The problem is NOT the DUI or the bad check. The problem is NOT the lost job, the lies, the thefts, or the criminal behavior. The problem is <u>the changed brain</u>. Focus on the root cause of the problem— the brain. Treat the brain and the rest of the body—and avoid focusing on the symptoms. The actions of the addict may cause real and lasting consequences (including bankruptcy and jail time); but, unless you address the root cause of the problems—the brain—the problems will persist, repeat, and get worse.

If you recognize that the problem lies in the brain, then you will begin to treat the root cause of addiction rather than the symptoms. And, if you treat the root cause of addiction, your actions will have a greater probability of producing positive results. Note that we did <u>not</u> state that your actions will produce positive results—there is no "cure", no magic treatment plan, and no medicine that cures addiction. Realistically, if you effectively treat the root cause of addiction, you can improve the probability of positive results but you can**not** guarantee positive results.

<u>Before Proceeding—Pause and Assess.</u>

Before we proceed, ask yourself if you are living through a challenging time, if you are tired, and if you are confused. If you are reading this book for when your life is relatively calm and settled, you will comprehend and retain more information than if you are reading the book at a time when your life is filled with large challenges. If you are living through that challenging time, do not stop reading—but do understand that you may comprehend and retain less information than you would otherwise comprehend and retain. This book is designed for you to read (and re-read) in one or two days. But, if you are reading it during a challenging time of your life, you will probably need to read and re-read various portions of the book. It's OK.

Moreover, much of the information in this book becomes clearer and more usable over a span of time. Therefore, even though you comprehend the information set forth in this book, it may take some time for you to understand how to apply the information. Take your time—adapt your use of this book to meet your needs.

Consider the following example. One of the leading drug treatment programs in the United States provides a week-long family seminar for families of addicts, and a female executive took the seminar. But, she did NOT *begin* to comprehend the information that the program provided to her until the third or fourth day, and she did not understand how to apply the information for several months. This female executive is very bright, but she wasn't able to comprehend, retain and use the information until she gained some normalcy in her life AND some experience with the illness of addiction.

This book will provide you with information that seems counter-intuitive, and it may be hard to fully understand the information. Nevertheless, the information will be easier to process, understand and use as normalcy re-enters your life, and as you acquire some experience managing the disease of addiction.

The Choice.

Now, let's resume. The brain is the source of the problem, and if you focus on treating the brain, you should produce better results for yourself. *Is this the answer? If I take this suggestion, will my addict recover?* No, today, the simple fact is that most addicts will die of causes related to their disease. Accordingly, this book seeks to explain addiction and to provide you with suggestions on how to improve *your* life. Today, there is no cure for addiction, and no treatment or medicine to ensure that the addict will seek, obtain, and maintain recovery. Therefore, we will provide you with some basic facts that can improve your understanding of the disease, and with lessons and skills that you can use to guide your decisions and actions. **You canNOT choose for the addict**, but you can choose how you deal with the disease.

Since most addicts die of causes related to their disease, we want to provide you with information and suggestions that <u>you</u> can use to guide <u>your</u> decisions and actions, and to empower <u>you</u> to obtain some normalcy in <u>your</u> life. You cannot force, cajole or entice the addict to seek and maintain their recovery—that power rests <u>solely</u> with the addict. But, if you learn the difference between helping and enabling, and you develop and work your own recovery program, you will increase <u>your</u> chances of living a more normal life and finding some serenity. You may not be able to control the addict; but with information, skills, lessons, and practice, <u>you</u> can choose to control <u>your</u> behavior and create a more normal and better life for <u>yourself</u>. <u>It is **your** choice.</u>

<u>Your recovery program</u> may cause the addict to reassess their options. If the addict finds that their family and friends are seeking their own recovery, the addict may discover that they cannot easily manipulate the family and friends, AND the addict may run out of options, deal with the consequences of their actions, and consider recovery. The addict may choose to do this or choose to not do this—only the addict has the power to make and sustain the choice. It is the addict's choice—it is NOT your choice. Only the addict has the power to seek their recovery, and only you have the power to seek your recovery.

The Brain and Addiction

This section describes how drugs affect and change the brain.

The human brain is the most complex part of the human body.[1] All of our thoughts originate from within the brain. Our brain uses electro-chemical reactions to transmit messages, to create memories, and to produce thoughts; and, further, our brain uses our experiences to adjust our thoughts and to improve our judgment based upon the experiences.[2] The electro-chemical reactions produce messages and feelings that generate memories and lessons—and this process <u>changes</u> our brain.[3] That is, once you learn to count; you store the experience and the memory of how to add and subtract; and you are able to take action based upon that memory (such as counting the change from a retail purchase).

The electro-chemical reactions also produce feelings (including euphoria, happiness, sadness, and pain), and those feelings produce memories that change our brain. For example, once you take your first steps as a child, you produce feeling of accomplishment and happiness; and the feeling of accomplishment and happiness change the brain and enable you to take action based upon that memory (such as walking to your favorite restaurant). We use our memories to literally change our brains—the memories create physical changes to the brain that permit us to store and access the memories, and to take future action.

We use our memories to create skills, such as the ability to compute (i.e. add, subtract, multiply and divide) and run (i.e. balance, walk, run, and sprint). Humans are not born with the capacity to compute and sprint; but, our brain stores and uses memories in order to allow us to produce thoughts, lessons and skills that affect our actions. Our brain stores our actions, experiences, memories, lessons and skills, and our brain uses them to produce <u>thoughts that guide our actions</u>.[4] In short, our memories <u>change</u> the brain so that the brain can produce useful thoughts that lead to useful action.

The brain also regulates the basic functions of the body, and it enables the person to interpret and respond to the person's environment. The brain manages blood flow, heart rate, breathing, and other basic body functions, and it can change these basic functions based upon the environment. The brain's ability to interpret experiences and to shape the person's memories, thoughts, lessons, and actions enables people to dynamically alter body functions in response to different situations. For example, in response to a building fire, the brain produces feelings of alertness and fear, and it directs the person to seek a safer environment. However, with proper training and experience, some people learn to enter a burning building in order to save lives and to prevent further property damage. The brain allows firefighters to deal with fires while simultaneously regulating their heart rate, breathing, muscles and emotion. The brain uses memories to manage a person's thoughts, actions, and the body's basic functions.

Therefore, as a general matter, the brain performs its functions by collecting and transmitting messages throughout the entire body though electro-chemical processes, and by changing the brain so that it can guide future actions.[5] Normally, natural chemicals (which are already in the body) initiate electro-chemical reactions which change the brain's structure, the memories that are stored, the thoughts that are produced, and the actions that are taken.[6] The structure and the "wiring" of the brain change according to the parts of the brain that are affected by the chemicals; and the chemicals may affect some parts of the brain more than other parts. Over time, the presence of the chemicals, and the strength of the chemicals, will change the brain and the thoughts that it produces, and the person will take action based upon the thoughts produced by their brain.[7] Normally, the release of the natural chemicals allows people to store useful memories, to produce useful thoughts, and to produce useful actions; however, when the drugs release these chemicals, the memories produce harmful thoughts and harmful actions.

Drugs produce the same or similar chemicals to those produced by the brain (or cause the brain to release its own chemicals), and the drugs change/re-wire the brain to store altered memories, to produce inaccurate thoughts, and to take harmful actions.[8] The drugs produce or mimic the brain's natural chemicals that produce euphoria, pleasurable moods, and confidence, or the drugs cause the brain to release its own chemicals to produce euphoria, pleasurable moods, and confidence.[9] Drugs cause the release of dopamine into the brain in order to produce euphoria; or serotonin in order to produce pleasurable moods; or norepinephrine in order to produce confidence and lower inhibitions.[10] Dopamine, serotonin and norepinephrine (also known as neurotransmitters) are chemicals that are naturally found in the brain—they are a natural part of the human body.[11] But, the chemicals produced or released by drugs are <u>much stronger</u> than the chemicals produced by the body, and therefore the "experiences," "feelings", "memories" and "lessons" processed by the brain are much stronger and more intense than normal experiences, feelings, memories and lessons.[12] Drugs can release two to ten times the amount of chemicals (such as dopamine, serotonin and norepinephrine) than the brain can naturally create to produce pleasure; and, some drugs can produce the pleasure for a much longer period of time.[13] The drugs therefore produce more impactful experiences, feelings, memories and lessons. If the drug user was feeling depression, anxiousness, or pain, the drug quickly and effectively took away the depression, anxiousness or pain. That is, through the use of drugs, the drug user found a way to "control" their bad feelings of depression, anxiousness or pain and to produce good feelings of euphoria, pleasurable moods, and confidence.

The drug **solved** the drug user's depression, anxious, or pain AND it stored a strong experience, memory and lesson related to the use of the drug. **The drug was the solution**. The chemicals released by the drugs will "change/re-wire" the brain, and the brain will thereafter produce thoughts that value and reward the use of drugs. The drug user's "free will" and "rational thought process" are minimized by the changed brain in favor of thoughts that value and reward drug use because the drugs "solved" the drug user's problems.

Abusive drug use produces thoughts that value and reward the use of drugs, and it affects the body in other ways as well. Abusive drug use can cause the body to create a tolerance to the drugs.[14] The brain adjusts to the overwhelming surges in the chemicals and often produces less natural chemicals (because the chemicals are externally provided) and by reducing the number of brain receptors (to avoid overloading the brain's receptors).[15] At the same time, in order to feel the euphoria produced by the overwhelming surges of chemicals produced by drugs, the brain produces thoughts that drive the addict to take larger amounts of drugs. As a result, the brain attributes less reward and less pleasure to the natural chemicals that are naturally produced by normal pleasures and experiences (such as a good meal, love and affection, a smile, or companionship)—and the addict is driven by their brain to seek and use more drugs to obtain the same effect. Abusive drug use produces thoughts that value and reward the use of drugs, and it devalues the pleasures and rewards of normal activities.

Drugs alter the experiences and memories produced by the brain AND the physical condition of the brain AND the manner in which it produces thoughts AND the thoughts that are produced. The changes to the brain affect the addict's memories, thoughts, lessons, skills, emotions, perceptions, predispositions, and actions. Accordingly, whereas a normal person is directed by their brain and their memories to produce useful thoughts that lead to useful action, the addict is <u>directed by their changed brain and memories</u> to take actions that are perceived as useful and rational by the addict (and harmful and irrational by us). But, recall that the drugs provided the drug user with a way to "control" their feelings of depression, anxiousness or pain and to produce euphoria, pleasurable moods, and confidence—and they therefore value the drugs. Therefore, when you ask the addict, "Can't you see the harm that you are doing to yourself?", the addict literally cannot "see" or perceive the harm because their brain produces thoughts that "see" and "perceive" their actions as useful and rational (whereas we see the actions as harmful and irrational). The addict's changed brain produces thoughts that perpetuate the disease.

To further understand how drugs affect the addict's thoughts and actions, it is useful to understand the basic structure of the brain, its basic functions, and how it prioritizes and orders thoughts. The three structural areas of the brain that are affected by drug abuse are the **brain stem**, the **limbic system,** and the **cerebral cortex**.[16] The **brain stem** controls the basic functions of the body.[17] These basic functions include the body's heart rate, breathing, and sleeping rhythm. The brain stem manages the basic functions needed to sustain life. The **limbic system** controls the brain's reward functions.[18] The limbic system controls and regulates the body's ability to feel pleasure, pain, and reward, and to feel positive and negative emotions.[19] The positive and negative emotions produced by the limbic system direct the individual to either repeat OR to discontinue actions and behaviors.[20] Normally, these feelings and emotions motivate the individual to repeat behaviors that are critical to the person's existence and survival, such as eating, sleeping and reproduction. The **cerebral cortex** controls the "thinking" part of the brain.[21] The cerebral cortex is responsible for the individual's ability to think, plan, address issues, make decisions, and solve problems. The three areas of the brain control the basic functions of the body, the person's perceptions of pleasure and pain that guide future action, and the ability to "think" and make informed judgments.

Exhibit A: Parts of the Brain (Cerebral-Cortex, Limbic System, and Brain Stem)

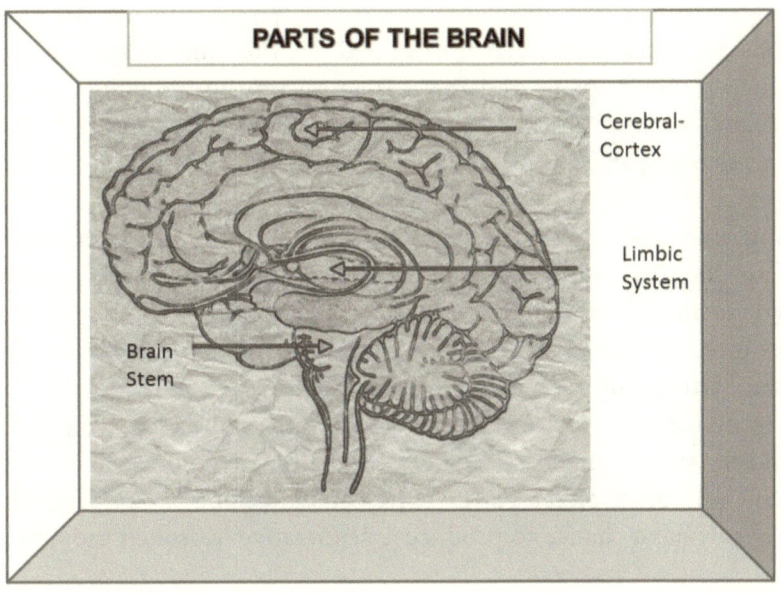

Memories operate in a hierarchical order, and memories tied to the limbic system have higher priority over memories tied to the cerebral-cortex/neo-cortex.[22] Memories tied to the limbic system possess higher priority because they are used to preserve and maintain the basic functions of the body. Further, memories stored in the limbic system are processed 4 to 5 times more quickly than memories stored in the neo-cortex.[23] As a result, memories tied to the limbic system are or become "instincts" and require minimal conscious or deliberative thought. These memories—which can behave like instincts—allow people to quickly act in order to preserve life. For example, if a speeding car rushes in your direction, your brain accesses memories and produces thoughts that cause you to "get out of the way"—and you "instinctively" take action rather than pondering the merits of taking action. In a different situation, if you are a trained police officer and you are confronted by an armed suspect, your training will cause you to "instinctively" raise your weapon to preserve public safety and your life. In people, memories in the limbic system—instinctual actions—occur before the person can take deliberative action.

Memories tied to the limbic system are automatic/reactive, instinctual memories. These memories control and regulate the person's ability to feel pleasure, pain, and reward. Further, <u>and very importantly</u>, these memories direct the individual to either repeat OR to discontinue actions and behaviors. Memories tied to the limbic system (which produce pleasure, pain, and reward) motivate the individual to repeat behaviors that are critical to the person's existence and survival, such as eating, sleeping and reproduction.

Strong emotional memories that are stored in the limbic system and tied to "reward" and "survival" are stronger memories. As previously stated, memories stored in the limbic system will be processed 4 to 5 times more quickly than memories stored in the neo-cortex. Drugs that produce strong euphoria and that store the memory in the limbic system will more forcefully drive thoughts and actions <u>more effectively</u> than the memories stored in the neo-cortex. Therefore, memories in the limbic system will be stronger AND drive thoughts and actions <u>before</u> the neo-cortex has a chance to produce a deliberative, reasoned thought.

As previously described, the chemicals released by the drugs will "change/re-wire" the brain to value and reward the use of drugs, <u>and</u> they affect different parts of the brain. Drugs have a pronounced effect upon the limbic system (which controls and regulates the body's ability to feel pleasure, pain, and reward, and to feel positive and negative emotions).[24] Memories of drug use cause the person to value and reward the continuation of drug use, <u>and</u> the memories in the limbic system link that value and reward to survival.[25] Drugs use the brain's chemicals and its structure to "hijack" the brain.[26]

In the hijacked brain, drugs become the primary source of pleasure and drug use is perceived as an act of survival—drugs alter the experiences and memories produced by the brain AND the physical condition of the brain AND the manner in which it produces thoughts AND the thoughts that are produced. As the drug user becomes an addict, drugs change the parts of the brain that control pleasure and reward and that direct the person to repeat behaviors, and the drugs cause the addict to react more quickly than they can produce deliberative or reasoned thoughts. The drugs change the brain—which we use to guide our everyday lives—to reward and perpetuate drug use. As the drug user becomes an addict, the changes caused by addiction will "hijack" the brain and perpetuate drug use.

The hijacking of the brain by drugs occurs in a predictable, biological and scientific manner. Drugs produce extreme pleasure through the release of chemicals within the brain, and this has a direct effect upon the brain's limbic system and its control and regulation of the brain's reward functions. The positive and negative emotions produced by the limbic system direct the individual to either repeat OR to discontinue the actions and behaviors. Normally, pleasure is initiated by an experience and a memory. The extent of the pleasure will vary between individuals; however, fairly consistently, the brain learns to create memories from experiences linked to pleasure and it creates lessons that reward these behaviors with feelings of pleasure. In the case of drug addicts, drugs become the primary source of pleasure and drug use is perceived as an act of survival, which is repeated, and repeated, and repeated.

As you have read, a brain that is affected by drugs will change and it will process data differently than a normal brain. Therefore, when the addict is confronted with negative consequences arising from drug use, their brain (which is changed) will produce thoughts that are different than the thoughts of a normal brain. It is difficult for the addict to stop using drugs because their brain <u>instinctively</u> views drug use as <u>critical</u> to "normal" life, as a <u>solution to negative feelings</u> (such as depression, anxiousness or pain), as <u>source of pleasure</u>, <u>and</u> as a <u>means of survival</u>. As a result, the brain produces thoughts that value drug use more than normal life experiences, and the brain minimizes or ignores the negative consequences of drug use. Drug use changes the brain and the thoughts produced by the brain.

Sustained drug use worsens the problem. Sustained drug use changes the brain and causes it to produce thoughts that value and reward drug use, and it diminishes the brain's ability to produce natural chemicals.[27] As a result, stopping drug use creates a wide gap in the amount of chemicals needed by the addict's brain to produce pleasure, good mood and confidence, and the brain's natural ability to produce or process the chemicals. The drug user perceives the lack of the chemicals, and the drop in pleasure, good feelings and confidence, with drug abstinence.[28] At least initially, the brain produces thoughts that associate abstinence with the lack of pleasure, good feelings and confidence, and the source of pain.[29] The brain produces thoughts that encourage the addict to correct the bad feelings, anxiousness, discomfort, and pain with renewed drug use. Nevertheless, if the addict remains sober, the brain will begin to produce natural levels of chemicals, and the addict can begin to experience normal pleasure, good feelings, and confidence, and the pain will subside.[30]

Stop and take note: First, the brain is structured to learn from positive and negative experiences and memories, and to repeat those activities that are "perceived" as positive and that produce pleasure, reward and continued life. The brain records memories, and it teaches us to repeat the experiences and behaviors that it perceives are positive. Drugs utilize this same process to teach the brain to produce thoughts that perceive drug use as a positive behavior. Drugs utilize this same process to compel the addict to use drugs in order to obtain pleasure, reward and continued life, and to minimize or disregard the ill-effects of drug use.

Second, the addict's brain will produce thoughts that we see as illogical and irrational. But, to the addict, the thoughts are logical and rational, and the thoughts support continued drug use. If you provide the addict with a logical basis why they should stop their drug use, your logical statements will be processed by the addict's brain—and the addict's brain will produce illogical thoughts and conclusions. Your logic—no matter how profound or compelling—will not cause the addict to become and stay sober. The addict's brain—not your brain—must find the rationale and the motivation to seek recovery and to stay sober.

Third, the brain produces thoughts that lead the addict to take actions that the addict's brain perceives are correct, but which we perceive as evidence of a lack of self-control and poor decision-making. Actually, the addict is behaving as he or she is instructed by their brain. Moreover, and very importantly, the addict's brain recalls that the addict used drugs as a mechanism to cope with life's challenges and problems, and that the drugs insulated the addict from their challenges and problems for some period of time. The addict's brain is taught to process thoughts that link drug use with *solving* the addict's challenges and problems *rather than* being the source of the addict's challenges and problems. In case of an addict, the brain's physical structure and functioning have been "hijacked" by the drugs to produce thoughts that lead to continued drug use. The brain may lead the addict to conclude that continued drug use is highly desirable because any negative consequences are minimal in comparison to the euphoria, reward, pleasure, lack of pain, and "results" produced by the drugs. In effect, the brain tells the addict to ignore the negative consequences because the costs are minimal in relation to the benefits produced by the drugs (i.e. the euphoria, reward, pleasure, lack of pain and "results").

Addiction is a brain disease[31], and it produces real and profound changes in the brain and the thoughts produced by the brain. Addiction is NOT a moral affliction. Addiction is NOT an indicator of intelligence. Addiction is NOT a criminal offense. Addiction is brain disease, and the disease should be treated as a disease. Treat the disease, and the symptoms can improve. Treat the disease rather than the symptoms.

Addiction is NOT a Choice

The decision to use drugs was a conscious choice. The choice may have been made when the individual was a teenager, when the individual was under stress, or when the individual was under the care of a doctor and taking prescribed drugs to reduce pain. Although the choice to use drugs was volitional, the person's change from drug user to drug addict occurs biologically. There is no conscious choice by the drug user to become a drug addict. The change from drug user to drug addict is made by the body—it is not a choice.[32]

Some factors increase the likelihood that a drug user will become a drug addict. One of those factors is genetics. Research indicates that a significant portion of the risk factors that lead to addiction can be traced to the individual's genetics.[33] For example, children who have one alcoholic parent have a 34% greater likelihood of becoming an alcoholic as compared to a child of nonalcoholic parents.[34] Children whose parents are both alcoholics have a 400% greater likelihood of becoming an alcoholic as compared to a child of nonalcoholic parents.[35] Children whose parents are both alcoholics, and who have a grandfather who is an alcoholic, have a 900% greater likelihood of becoming an alcoholic as compared to a child of nonalcoholic parents and nonalcoholic grandfathers.[36] Children of alcoholic parents are generally less sensitive to the sedative effects of alcohol, and therefore they can "hold their liquor" better than other people. Children of alcoholic parents may also be more sensitive to the pleasurable effects of alcohol, and they find alcohol consumption to be more rewarding than other people. Both of these genetic characteristics can lead to increased alcohol consumption, and these genetic characteristics can lead to quicker transition from alcohol use to alcohol addiction. Therefore, while one individual may be able to engage in alcohol consumption and not become an addict, a different individual may engage in the **same** behavior for the **same** period of time *and move from an alcohol user to alcohol addict*. The same behaviors by different people can lead to different results—one may become an addict, and the other may not become an addict even though they engaged in the same behaviors.

Fear and Hope

If the drug user becomes a drug addict, they will remain an active drug addict until they choose to seek drug recovery. Many factors will affect this choice. However, there are two basic factors that usually lead to recovery: fear and hope.[37]

Exhibit B: Fear and Hope

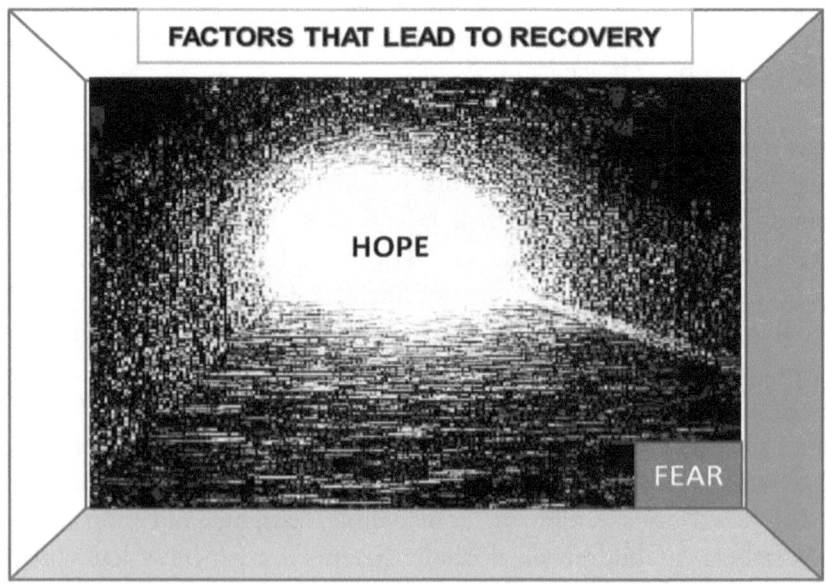

FACTORS THAT LEAD TO RECOVERY

HOPE

FEAR

The addict may envision the possibility of drug recovery when they fear the prospect of remaining an addict more than they fear the prospect (and risk) of change.[38] This event may occur when the addict experiences something that threatens their basic value system. The event can be a near-death experience or the use of "bad" contaminated drugs. The event can occur from prolonged hunger or the stress of living on the streets. The event can also occur in response to repeated muggings or rapes. When the fear of remaining an addict and continuing this lifestyle challenges the addict's basic value system, they may begin to envision the possible need for drug recovery. The moment when the addict's basic value system becomes fundamentally threatened is often referred to as "hitting bottom".[39] However, it's important to understand that "hitting bottom" is not required in all cases AND that "hitting bottom" can be the addict's death. Many addicts will die before they seek recovery.

The addict may also envision the possible need for recovery when a sense of hope is reawakened in their consciousness.[40] The presence of hope may occur when they see a friend who achieved some success with sobriety even though that friend was "worse off" than the addict. Hope may also arise when the addict has no means to buy additional drugs, they go without drugs for several days, and they survive the withdrawal symptoms that they believed were not survivable. Hope may also arise from attending an Alcoholics Anonymous (AA) meeting and developing a friendship or connection with another person who attends the meeting. There is no single event that will spark hope in all addicts; however, if the addict can envision hope, they can foresee the possibility of recovery. Do not prescribe the path to recovery. Instead, offer examples and observations, and permit the addict to use their own hope and fear to create their own path to recovery.

Do not search for the magic words that will instill fear or hope into the mind of the addict—there are no magic words. We have not heard of a single addict who is "saved" because their loved one told them to get better. Addicts get better because they want to get better, and not because we want them to get better. Therefore, rather than trying to find the magic words, have a conversation with the addict and explore their feelings. Listen carefully—rather than preach—and perhaps the addict will mention or describe their fear and hope. Assist the addict to use their fear and hope to envision the possibility of recovery.

Consider the following: the addict cannot provide us with the magic words that will allow us to have "hope". Therefore, it is logical that we cannot provide the magic words that will allow the addict to have "hope" or "fear". Hope and fear must come from within the person who will use those feelings to seek recovery—this is true for the addict, and for us too.

Barriers to Recovery

Even if the addict can envision recovery, there are several factors that will impede their decision to seek recovery. Now that the brain is changed, and the thoughts produced by the brain are also changed and the addict's choices are clouded by at least five factors: (1) withdrawal, (2) age, (3) other brain illnesses, (4) friends, and (5) family.[41] These factors should not prevent the addict and their family from taking action; however, these factors should be considered because they will affect the addict's thoughts, perceptions, and willingness to seek recovery.

Exhibit C: Barriers to Recovery

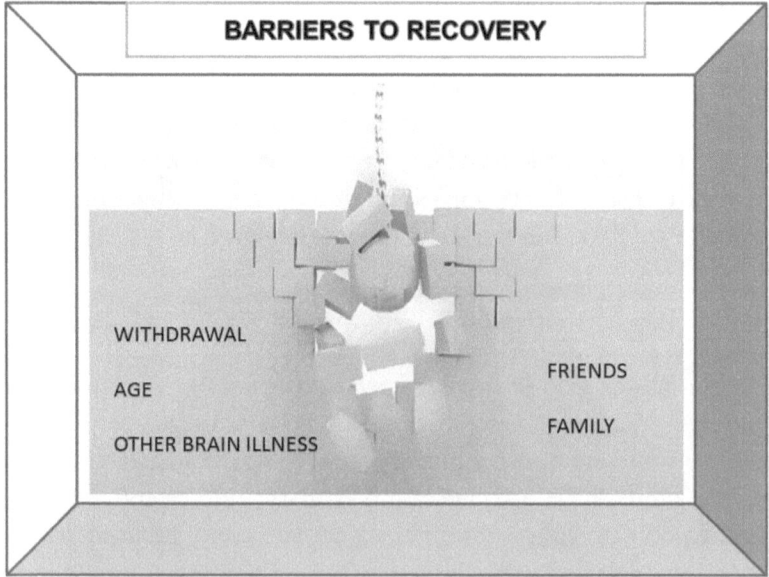

The first factor is withdrawal. The addict's ability to make informed decisions is affected by drug withdrawal. The addict will experience two stages of withdrawal: acute withdrawal and post-acute withdrawal.[42] Acute withdrawal begins immediately after drug use is discontinued. The effects of acute withdrawal will vary based upon the drug; however, most drugs produce significant physical and emotional symptoms.[43] These symptoms can include pain and discomfort. In regard to physical withdrawal, the addict may experience sweating, heart palpitations, muscle tension, breathing difficulty, tremors, nausea and diarrhea. Often, acute withdrawal feels like a very bad flu, with symptoms such as seizures, sweating, goose bumps, vomiting, diarrhea and tremors.[44] Acute withdrawal can be felt and perceived by the addict as a struggle for survival. In regard to emotional withdrawal, the addict may feel anxiety, restlessness, irritability, headaches, poor concentration, insomnia and depression. In general, drug withdrawal causes symptoms that are <u>opposite</u> to the effects produced by the drugs. [45]

When the addict stops using drugs, the brain attempts to rebound and to recover normality—and this rebound will produce the symptoms that are described above.[46] These symptoms are normal and should be anticipated. It is advisable to seek medical assistance in treating the withdrawal symptoms, and new medicines can lessen the discomfort associated with withdrawal. Withdrawal from opiates can be extremely uncomfortable, but the withdrawal from opiates is not generally life-threatening.[47] However, withdrawal from alcohol and depressants can produce very dangerous physical withdrawal symptoms (including heart attacks, strokes and seizures).[48] It is especially important to seek medical treatment when dealing with withdrawal from alcohol and depressants.[49]

Acute withdrawal will last several weeks; however, the most severe symptoms will be experienced during the initial two or three weeks.[50] Acute withdrawal is followed by post-acute withdrawal.[51] During this second stage of withdrawal, the addict will have fewer physical symptoms, but they will experience more emotional and psychological withdrawal symptoms.

As previously stated, withdrawal (including post-acute withdrawal) occurs because the brain is attempting to return to its normal functioning. As the brain improves it functions, the level of brain chemicals will fluctuate as the brain seeks to achieve new chemical equilibriums. Post-acute withdrawal symptoms include anxiety, irritability, variable energy levels, lack of enthusiasm, poor concentration, insomnia, and mood swings.[52] Post-acute withdrawal resembles a roller coaster of emotions, which can last 6 to 18 months (and sometimes longer).[53] The addict should anticipate the roller coaster of emotions, and they should understand that the ups and downs of the "roller coaster experience" will lessen over time. The addict should not hopelessly despair—they should hang on because it will get better. As the addict refrains from drug use and the brain begins to heal, the episodes will become less frequent and less severe.[54] Post-acute withdrawal usually lasts 6 to 18 months.[55]

Withdrawal affects the addict's emotions and brain chemistry, and therefore be aware that withdrawal affects the addict's ability to make informed decisions.

The second factor affecting the addict's ability to make rational choices is age. We expect teenagers to make rash and uninformed decisions—and this is no different for teenage addicts. The brain is not fully developed in males until age 25; and, since the brain is not fully developed, it is easy to understand why teenagers (and teenage addicts) make decisions and take actions based upon incomplete information.[56] Teenagers favor memories and actions that produce pleasure and reward, whereas mature adults have and use memories that empower them to better balance their need for pleasure, reward and action. When you consider that brains are not fully formed until the age of 25, the risky acts of teenagers are biologically predictable. It is also easy to understand why many teenagers (and teenage addicts) possess poor decision-making skills and poor judgment.

In addition, recent research indicates that two neural and psychological systems interact with each other as teenagers progress toward adulthood.[57] These systems have a pronounced effect upon teenage behavior as compared to adult behavior. The first of the systems pertains to emotion and motivation. The first system is connected to the biological and chemical changes related to puberty, and it involves the areas of the brain that respond to rewards. Research indicates that teenagers are not reckless because they under estimate the risks—they are reckless because they place an excessive emphasis on rewards and the part of the brain that blocks risk-taking behavior isn't fully developed until age 25.[58] The reward systems of the teenage brain are much more active than in the adult brain. This emphasis on rewards combines with the second system: peer approval. Teenagers are very concerned with their social circles and the opinions of their friends. Teenagers receive social rewards from their peer groups, and they are more likely than adults to seek peer approval. The reward system of the teenage brain is more active when they are in social settings with peers and friends. Accordingly, be aware that teenagers are predisposed to seek (a) rewards and (b) peer approval even if their actions require them to engage in risky behavior. These predispositions should be considered and addressed in order to encourage the teenager to change their behavior.

In addition to partial development of the brain prior to age 25 and the predispositions based upon age, it is important to understand that drug use affects the development of the brain; and that the brains of teenage addicts develop more slowly than the brains of "normal" teenagers.[59] In general, the maturity of the addict is "stuck" at the age that they began taking drugs, and maturity development does not appreciably progress until drug use is stopped.[60] When drug use is discontinued, the addict will make choices based upon the development of their brain at or near the age that drug use began. Therefore, if the addict begins drug use at the age of 16 years and stops drug use at the age of 25 years, the addict may frequently make the decisions of a 16 year-old. This is normal—account for it. Provided the drug use did not cause permanent damage, the brain can heal, and addict's maturity, coping skills, and judgment can progress.

Age affects the ability of the addict to make informed decisions.

The third factor is the presence of other brain illnesses. At this point, remember—the brain is an integral part of the body and it deserves medical treatment. Set aside any and all stigmas, prejudices, and phobias that you may possess regarding brain illness. Unless you consciously set aside those stigmas, prejudices, and phobias, you may lessen the addict's and your chances of recovery. <u>Brain illness exists, the brain is a part of the body, and the brain deserves medical treatment</u>.

Persons with brain illnesses are generally more likely to engage in risky behaviors that lead to addiction.[61] Persons with brain illnesses may be less inhibited and may be more willing to take risks. They may also be less able to assess the nature of the risk being taken or the consequences of their actions, and these risky behaviors can lead to drug use and addiction. Further, some brain illnesses (such as anxiety, depression, bipolar disorder, and schizophrenia) have strong correlations with addiction. Addicts may find that the symptoms of these illnesses are lessened through drug use, and drug use becomes a form of self-medication. You should understand that the symptoms of drug abuse and other brain illnesses sometimes overlap, and that drug use sometimes accentuates <u>or</u> masks the symptoms. The overlap can make diagnosis and treatment more difficult.[62] However, to be effective, the addict should receive treatment for all diseases. Since the addict may have different brain diseases, all the diseases need treatment to improve the chances of recovery.

Seek medical treatment. The presence of a brain disease will affect the addict's judgment and decision-making abilities. Therefore, to improve the addict's judgment and decision-making, it is important to seek medical treatment for all brain diseases. Seek medical treatment from physicians who treat addiction and brain diseases; and, understand that you may need to find a different physician for each disease. Properly treating the diseases will affect the addict's choices, and these choices will affect their recovery.

All of the addict's <u>brain illnesses</u> should be treated (and the non-brain illnesses should be treated too).

The fourth factor is friends. Friends influence the addict's thoughts and behavior—especially when the addict is a teenager. People form social relationships with other people with whom they share the same interests, values, and outlooks. Since the addict is usually a risk taker, they will form friendships with other individuals who share the same risk-taking predispositions. Since people often look to their friends for affirmation and support, friends will frequently take actions together. If those actions involve drug use, the addict will find acceptance and support for drug use among their friends.

Drug use among friends can become the primary behavior upon which the friendship revolves.[63] The friends may recount their experiences with drugs, and compare and contrast their experiences and choices of drugs. This recounting and shared experience may lead the friends to experiment with different drugs and different doses of drugs. The friends may discover that their initial choice to take drugs was a voluntary decision, but they may later realize (as individuals) that they are finding it more difficult to control their drug use. Nevertheless, in order to maintain the friendship, the individual drug users (friends) may hide or discount their difficulty in controlling their drug use. Indeed, to maintain the friendship, the friends may individually take additional risks that lead to drug addiction in order to keep one or more of the friends.

With regard to friends and drug use, peer pressure influences people, and leads them to do things that may be resistant to do or might not otherwise choose to do. Peer pressure also influences people to do things more often and with greater intensity than they might otherwise choose to do without the pressure. In regard to teenagers, the influence of peer pressure is heightened because they are at a stage of development when they are separating from their parents' influence, but they have not yet established their own values or an understanding of the consequences of their behavior. Teenagers are looking for social acceptance, and they are generally more willing to engage in risky behaviors in order to gain and retain social acceptance. The influence of friends is an important factor to consider because this influence will affect the addict's choices.

There is an old adage that states, "Show me your friends, and I will show you who you are"—and this is especially true in regard to the addict. This old adage is a reliable predictor as to the thoughts and behaviors of the individuals who are friends. Therefore, if you see drug use among the addict's friends, there is a substantial likelihood that the addict is using the same drugs. There is also a strong likelihood that the friends support the drug use, and that the friendship is actually supporting and promoting ongoing drug use. Since drug use will affect the thoughts that the brain produces, when a group of friends engage in drug use, their brains will produce similar thoughts that reward and affirm continued drug use by all of the friends. Further, the friends will collectively affirm and reward continued drug use, and the individuals may discount or reject thoughts which disfavor drug use. As a result, the addict will find it increasingly difficult to question their drug use.

Notwithstanding the foregoing, friends can also drive and support sobriety. If the friends discover that one of them has a drug problem, they can encourage that friend to seek help. Or, if the addict wants to stop their drug use, they may need to find a new circle of friends. In both cases, the same "friend" influences that can support drug use can be used to support sobriety.

Friends influence the addict's thoughts and behavior—therefore, the addict should seek friends who support sobriety.

The fifth factor is family. Drug use does not have a single cause or trigger, but drug use and behavior problems often stem from family indicators. The family indicators include genetics, drugs used by other family members, attitudes regarding drug use, inconsistent parenting behaviors, and enabling behaviors. For these reasons, it is very important that the family act in the unified way toward the addict.[64]

The family's ability to deal with codependency and enabling behavior can materially affect the health of the family and the addict. The addict needs help, and a common practice is for one family member to seek to provide "help" by "enabling" the addict to avoid the negative consequences of their actions. The family member may see the enabling behavior as a means to assist the addict to more carefully assess the consequences of their actions and to avoid repeating those actions. In this regard, enabling behavior includes covering up addiction-related problems at school or in the workplace, making excuses for inappropriate behaviors, or paying the costs or expenses incurred because of the drug use. Although the addict's behaviors are viewed as negative by the family member, the family member permits the addict's behavior in order to "protect" the addict or to give them "another chance". These family members may become obsessed with trying to control the addict's behavior, and this obsession becomes a "codependency".

When the addict is a teenager, the presence of enabling behaviors and codependency is often confused with the parent's desire to provide instruction, food, and shelter for their child. Indeed, since the parent understands that the teenage addict will make "adolescent decisions" using a teenage brain, the parent may seek to provide the teenager with the time needed for the addict to mature and to make better decisions. However, enabling the teenage addict usually leads to continued drug use, and this further leads to stifled brain development, diminished maturation, and reliance on poor coping skills.

In order for the addict to get better, the family (as a group) should seek to get better.[65] If the family develops the skills that are needed to identify and deal with addiction, the family will develop the skills needed to seek <u>family recovery</u> regardless of whether the addict chooses to continue drug use or to seek recovery. To acquire recovery skills, the family should avail itself of support groups and/or family counselors. If individual family members need more assistance, they should seek help from doctors who treat addiction and from drug counselors. In many cases, when the family learns to set boundaries and to deal with the addiction with common skills, their family will be less susceptible to manipulation by the addict. As the family learns to not become obsessed with controlling the addict, the addict will begin to be confronted with the negative consequences of their addiction, and the addict may begin to associate these negative consequences with their drug use. This confrontation may result in the addict producing thoughts that conflict with the previous thoughts that favored continued drug use, and the addict may begin to think about the possibility of diminished drug use or stopping drug use.

Addiction is a <u>family</u> disease.[66] Acquiring an understanding of how the disease affects the family will improve the family's ability to make better-informed decisions.

You now know that addiction is a brain disease, and that the decision to seek recovery will be affected by many factors. If you keep this information in mind, you may be able to influence the addict to consider recovery. By addressing the brain disease and the five barriers to recovery, it is possible to provide an environment in which the addict decides that it is in their best interest to seek recovery. However, keep the following four items in mind. First, while some families are able to force the addict into a recovery program, most addicts succeed at recovery when they (the addict) make the choice to seek recovery. Second, there is no formula that you can follow in order to make the addict seek recovery. Use the information that you find in this book, seek professional assistance, and attempt to improve the probability that your addict will seek recovery. But, remember that you cannot force the addict's brain to produce the thoughts that you want it to produce. Only the addict has the ability to seek recovery and to stay in recovery. If you doubt this statement, then perform the following exercise. Think of a political issue that you strongly support and the politician who most disagrees with your position. Now think of the statements and arguments that you know **should** make the politician change their mind, and imagine that you provided these statements and arguments to the politician. Consider the probability that you would successfully change the mind of the politician. Hopefully you concluded that your chance of success is very remote—no matter your political outlook, you realistically stand very little chance of changing the politician's mind because they perceive and process the same facts very differently. The same is true with the addict—they perceive and process the same facts very differently. Third, recovery is not a linear process. The addict may begin a recovery program and then relapse. The addict may conclude that they need to seek recovery, and then decide that their addiction is not a problem. Addicts may, and often do, progress and regress in the recovery. Expect relapses to occur, and encourage the addict to restart their recovery program as soon as possible after a relapse. Fourth, remember that many addicts succumb to their disease. Stay hopeful, and realistic.

This book seeks to provide you with basic facts and information concerning addiction. Accordingly, you now understand that addiction is a brain disease, and that you cannot force the addict's brain to produce the same thoughts that a non-addicted brain produces. You also understand that addiction changed the addict's brain, and that the addict's brain produces thoughts that (1) encourage and reward continued drug use and (2) minimize or ignore the negative consequences of drug use. You also know the five factors that are barriers to recovery. You have a basic understanding of the disease of addiction, and some understanding of why the addict finds it very difficult to stop their drug use.

The following pages will introduce you to the stages of recovery, and provide you with some options regarding recovery programs.

The Decision to Seek Recovery: Stages of Recovery

There are many different models that describe the process of recovery, and the names of the stages of recovery may vary among the different models. We will describe a general model, but it is by no means the definitive recovery model.

Recovery generally happens in six stages, and the stages are pre-contemplation, contemplation, preparation, action, maintenance, and relapse[67]. If there is a relapse, recovery requires that the addict actively re-engage recovery at a previous stage. The addict seeking recovery may spend considerable time in one stage, and less time in another stage. Also, the addict may partially proceed to a higher stage and become stuck between two stages. Further, the addict may successfully proceed toward a higher stage, and then fall back one or more stages. Recovery rarely occurs in an orderly, serial and straight-line basis.[68] Hope for progress; prepare for setbacks; and seek ways to motivate the addict (and yourself) to keep focused on seeking recovery for your own reasons.

In the pre-contemplation stage, the addict will typically deny that they are an addict, and the addict will believe that their addiction is not a problem.[69] In this stage, the addict will typically believe that they can control their drug use or that their drug use produces no negative consequences. To assist you and the addict to determine whether the user has become an addict, you should consider utilizing various self-administered addiction tests. These tests (such as the CAGE and

Narcotics Anonymous tests) provide you with a questionnaire, and your answers to the questionnaire will provide you with scores that assess the probability of addiction[70]. You can find these tests on the Internet. The tests will pose questions such as:

- C: Have you ever thought about or attempted Cutting down on your drug use?
- A: Do you find yourself Annoyed when others criticize or comment on your drug use?
- G: Do you ever feel Guilty about your drug use?
- E: Do you ever need a drug as an Eye-Opener early in the day to get you going?

  If you scored 1, there is an 80% chance you have an addiction.
  If you scored 2, there is an 89% chance you have an addiction.
  If you scored 3, there is a 99% chance you have an addiction.
  If you scored 4, there is a 100% chance you have an addiction.

  and

- Do you drink or use drugs alone?
- Have you ever had a complete loss of memory as a result of drinking or drug use?
- Has your physician ever treated you for drinking or drug use?
- Do you drink or use drugs to build up your self-confidence?
- Have you ever been to a hospital or institution because of drinking or drug use?
- Do you lose time from work due to drinking or drug use?
- Is drinking or drug use making your home life unhappy?
- Do you drink or use drugs because you are shy with other people?
- Is drinking or drug use affecting you reputation?
- Have you gotten in financial difficulties as a result of drinking of drug use?
- Do you turn to lower companions and an inferior environment when drinking or using drugs?
- Does your drinking or drug use make you careless of your family's welfare?

- Has your ambition decreased since drinking or using drugs?
- Do you crave a drink of drugs at a definite time daily?
- Do you want a drink or drugs the next morning?
- Does drinking or using drugs cause you to have difficulty sleeping?
- Has your efficiency decreased since drinking or using drugs?
- Is drinking or using drugs jeopardizing your job or business?
- Do you drink or use drugs to escape from worries or troubles?

If you answer "Yes" to three or more questions, there is a substantial likelihood that your drug use constitutes drug abuse or drug addiction.

These are self-administered tests, and you need not show the results to anyone.

At the pre-contemplation stage, you may consider using an intervention to confront the addict regarding their drug use and the consequences of their drug use. In the pre-contemplation stage, the addict exhibits denial.

The second stage is contemplation, and this stage occurs when the addict recognizes that their drug use has become a problem and they begin to consider whether to seek help[71]. This stage can occur quickly or over a long period of time. Further, the addict may recognize that they have an addiction and need help, and then re-assess their situation and conclude that they do not have an addiction. In order for the addict to proceed through this stage, they need to become convinced that seeking recovery is better than continuing their drug use. The addict is more likely to seek recovery when the drug use leads to consequences that they believe are undesirable. These consequences may include losing a job; being arrested; exhausting available financial resources; or living on the street and experiencing hunger and violence. In this stage, family members should stress the positive aspects of recovery. Family members should assist the addict to see that recovery is in the addict's best interests— provide the basis for the addict to motivate themselves to seek recovery. The addict needs to see how recovery will benefit them—not you. In the addict's mind, it is all about them—not you. Family members should avoid enabling behaviors and permit the addict to feel (and deal with) the consequences of their drug use. In the contemplation stage, the addict considers the need for recovery.

The third stage is preparation. This stage occurs when the addict decides to change and to take steps to prepare for the change.[72] This stage is characterized by tentative progress. The addict has made the decision to change, however they may reverse that decision because they are still under the influence of drugs and they question the need for change. Nevertheless, it's important to keep the addict positive. It is important to support the addict's decision and to support the addict's perception of the benefits that they can obtain from their decision. It is very important to emphasize that the decision rests with the addict, and that the decision will benefit the addict. During this stage, the addict may make an appointment with a physician or recovery program, or they may plan to attend support group meetings. In the contemplation stage, the addict prepares to seek recovery.

The fourth stage is action, and this stage occurs when the addict actually takes <u>demonstrable</u> action to address their addiction.[73] The earlier stages are characterized by mental predispositions and decisions; but this stage is characterized by <u>demonstrable action</u>. During this stage, the addict may actually enter a recovery program, or actually keep their appointment with the doctor or attend a support group meeting. In this stage, they may undergo detoxification and begin a program. During the action stage, the addict begins to learn how to understand and manage their disease. The addict will begin to learn that the disease can be managed but not mastered—because the disease is cunning, baffling, powerful, <u>and patient</u>.

Addiction is a very real disease, but because it is a brain disease that is associated with negative symptoms and consequences, there is a societal prejudice against those persons who have the disease. Consequently, there is considerable disinformation about the disease, and the addict will need to unlearn this misinformation and begin to learn "real" information concerning the disease. This fourth stage begins the journey of recovery, and it begins the addict's and the family's education regarding addiction. Both the addict and the family should undertake the action stage with great vigor. The addict and the family will each have their respective recovery programs, and the greater the dedication and vigor with which they practice their respective programs, the better the probability of a positive outcome for the addict, the family, <u>**or**</u> both. The key element of this stage is <u>action</u>—this action should be focused, deliberate, and sustained. In the action stage, the addict takes action to seek recovery.

During the action stage, the addict will undertake their recovery treatment "program". Some addicts will take action, stop their drug use, and create their own recovery treatment program. However, it is advisable to seek professional help and to have the addict participate in a formal recovery treatment program. A formal recovery treatment program involves the use of knowledgeable personnel to treat the addict and to train the addict so the addict can understand the disease and <u>apply</u> their learned skills. <u>Good programs require the addict to *apply* their skills and provide the addict with feedback</u>. <u>Recovery requires action by the addict</u>.[74] Formal recovery treatment programs involve frequent and regular <u>participation</u> in support group meetings (such as Alcoholics Anonymous meetings). The formal recovery treatment program may also involve <u>working</u> a 12-step program with a sponsor, and meeting with addiction doctors, psychologists and psychiatrists to train the addict to <u>consciously change</u> their thoughts and behaviors—to take concrete steps to produce change. Utilize the resources that are available, and understand that the addict need not <u>and should not</u> try to develop their own recovery treatment program. Alcoholics Anonymous, the 12-Step program, and the sciences of Psychology, Sociology, and Addiction Medicine, already exist—the addict need not spend the time and research to reproduce them.

Addicts are very confident individuals, and they consistently overestimate their ability to produce desired results. The addict should resist their natural inclination to create their "own" recovery treatment program, and they should utilize the resources that exist. By utilizing existing programs, practices, skills, and medicines, the addict can improve their chances of recovery. The addict should use existing recovery programs, and adapt them to address the addict's specific needs.

The fifth stage is maintenance. The maintenance stage occurs when the addict takes action to initiate and sustain recovery and to avoid relapse. Maintenance begins after a sustained period of sobriety.[75] To begin a period of sobriety, many addicts enter into a recovery treatment program. A recovery treatment program usually lasts between 30 and 90 days, and it may be convenient to consider that maintenance begins at some period of time between 30 and 90 days. However, remember that the brain is affected by the effects of addiction for more than 90 days. Therefore, there is no magic number of days at which the addict begins the maintenance stage. Rather than assuming that the addict moved from the action stage to the maintenance stage, it is better to assume that the addict is in the action stage for a longer period of time so that they acquire the skills needed to sustain their recovery. In the maintenance stage, the addict takes action to sustain their recovery and avoid relapse.

People in the maintenance stage have usually ended their formal recovery program. In the maintenance stage, they typically rely on support groups such as Alcoholics Anonymous or Narcotics Anonymous, and they may go to counseling or therapy sessions. To successfully handle the maintenance stage, the addict should develop a strong set of skills and a proper understanding of the disease. To acquire this knowledge, the addict should work very hard in the action stage to acquire the required skills and knowledge, and the addict should not rush into the maintenance stage. Addicts are, by their nature, impatient and risk takers. Accordingly, we should encourage the addicts to work vigorously in the action stage to acquire, practice, test and temper their skills and knowledge so that they can successfully enter the maintenance stage.

It is very helpful to rigorously and consistently follow a recovery program—this will enable the addict to develop the skills and knowledge that they can use to sustain their recovery. The addict should be mindful that they should not become overly confident or complacent. In Alcoholics Anonymous, there is a slogan that addiction is "cunning, baffling and powerful".[76] Addiction is also "patient". During the maintenance stage, it is very important for addicts to recognize the cues and triggers that can lead to relapse. It is very important for the addicts to develop the skills and knowledge that are needed to proactively recognize the cues and triggers, and to take steps to either (a) avoid the queues and triggers or (b) address and minimize their impact. Since addiction is cunning, baffling, powerful, and patient, the addict must simultaneously move on with their life and vigorously work to maintain their sobriety program.

The sixth stage is relapse. Relapse begins when the addict mis-manages the warning signs that signal the progressive relapse process—where the process ends with alcohol or drug use.[77] The addict should recognize the early warning signs and address them before the process leads to alcohol or drug use. Nevertheless, expect one or more relapses to occur. This is normal. Do not automatically consider a relapse to be a failure. Recognize a relapse as a setback, and use it as a means for the addict to more strenuously develop the skills and knowledge required to sustain recovery. Relapses can occur for a period of one day, or for many years. There is no typical relapse or period of relapse. However, there are items that often lead to relapses. Relapses often occur when the addict is hungry, angry, lonely, tired, or anxious.[78] Therefore, the addict should recognize that these feelings may cause the relapse, and they should take prompt action to address feelings of hunger, anger, loneliness, tiredness or anxiousness. The more quickly that the addict sees the need to return to sobriety, and to address those things that caused the relapse, the more successful the addict will be in renewing their sobriety. A relapse can be a very good tool and reminder to the addict, and it can be used by the addict to prevent future relapses. Nevertheless, work very diligently to avoid relapses— because some relapses end with the death of the addict.

Relapse is not necessarily a sign of failure—it can be an opportunity for the addict to use their skills and knowledge in order to shorten the period of relapse and to restart their recovery. When an addict experiences a relapse, they may attempt to hide it from their family and friends. The addict may also attempt to hide it from themselves by minimizing the relapse and its effects. The addict may be ashamed or afraid to tell their counselor or their support group about the relapse, and they may believe that their families, counselors and support groups will view them as a failure. The stress of the relapse, and the fear that others will learn of the relapse, may prevent the addict from recognizing the relapse and taking action. Therefore, during the maintenance stage, everyone around the addict should recognize that relapses occur and commit themselves to having the addict quickly handle any relapse. A relapse should be taken seriously, and the addict should attempt to restart their recovery as soon as possible. If a relapse occurs, the addict may not be able to begin at the fourth stage (the action stage)—they may need to begin at stage two or three (the contemplation or preparation stages). No matter the stage where the addict begins, the important point is that the addict should be encouraged to restart their recovery.

There are some individuals and groups that believe that there is a stage called "termination", which occurs when the addict totally recovers from their addiction.[79] While this stage may occur for a small number of people, the vast majority of addicts will discover that addiction has permanently changed their brains and that they must manage the addiction for their entire lifetime.[80] Be aware that there is no pill, treatment, or program that can guarantee that the addict will totally recover from their addiction, and remember that many addicts will die from their addiction. Nevertheless, with treatment, and by following "their" program, many addicts can manage their disease and live productive lives. Therefore, if the goal is to provide the addict with a means to live a productive life, it is very clear that addicts should recognize that addiction is a lifetime disease, and that the addicts should acquire the skills and medical treatment necessary for them to manage the disease. Active use of these skills and medical treatment can actually empower the addict to create a purposeful life— probably with greater purpose and direction than before they acquired the addiction. With recovery, there is hope.

Brief Recap, and Pause

Let's briefly recap. Why can't the addict stop? They have a brain disease. Did the addict choose to become an addict? No, they chose to use a drug, and the use of that drug eventually caused the body to become addicted to the drug. That is, the drug user chose to use drugs, and the drug user became an addict when the body became afflicted by the disease of addiction. Are there barriers to recovery? Yes, those barriers include withdrawal, the age of the addict, the presence of other brain illnesses, friends, and family. If we deal with these barriers, there is a greater probability that the addict may choose recovery. Is there a medicine, treatment or program that will guarantee recovery? No, however there some generally predictable stages that describe the process of recovery. If the addict recognizes the stages of recovery and utilizes them, there is a greater probability that the addict will seek and maintain recovery. Well, what's next? We will next describe a basic framework for recovery.

Stop and consider what you just read. The information is very helpful if you remember it. You are trying to prevent the destruction of the addict's life, or perhaps struggling with merely keeping them alive. Often, families with addicts are so stressed and so busy dealing with the consequences of the addiction that they diminish their capacity to think calmly and clearly. This is very common—and normal. Take your time. Be patient with yourself and your family. Re-read this book and other resources, and have confidence that the facts and information you learn will become more clear as time passes and as you acquire more experience. The good news is that you are seeking objective facts and information—which places you in much better shape than most people who deal with addiction in their family.

The Recovery Triangle: A Framework for Recovery

To improve the chances of successfully using the stages of recovery, you should consider the three primary factors that affect addiction.

Three factors are largely responsible for drug use becoming drug addiction. Those factors are: (1) biology, (2) psychology, and (3) sociology.[81]

Exhibit D: Framework for Recovery

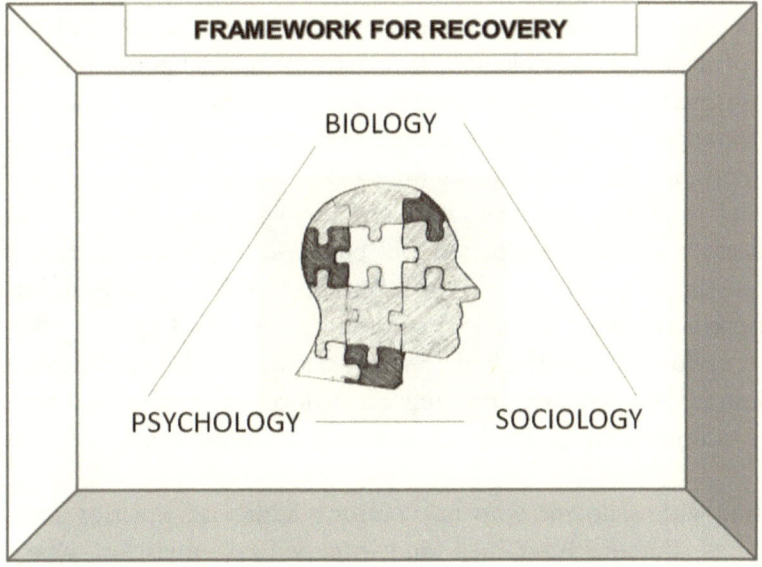

The biology factor describes the addict's possible biological predisposition to addiction, and the biological effect of the drugs upon the addict.

Statistically, certain biological and genetic factors are linked to addiction. According to the National Institute on Drug Abuse, scientists estimate that genetic factors account for between 50% to 60% (or more) of a person's vulnerability to addiction.[82] Similarly, as previously stated, research indicates that children who have one alcoholic parent have a greater probability of becoming an alcoholic themselves, when compared to a child whose parents are not alcoholics; and a child whose parents are both alcoholics have a 400% greater probability of becoming an alcoholic than a child whose parents are not alcoholics. Statistically, individuals who carry the genes of addicts have a substantially greater probability of becoming addicts. Therefore, even if science has not conclusively proven that there is a biological genetic predisposition to addiction, the facts show an individual with a family history of addiction has a statistically greater chance of becoming an addict.

Further, various biological factors influence a predisposition toward addiction. Alcoholics and their children can be less sensitive to the effects of alcohol, and they can therefore consume greater quantities of alcohol. The greater consumption of alcohol can cause the brain to require higher alcohol consumption in order to experience the pleasure produced by alcohol and this can lead to addiction. Further, in some alcoholics, the effect of alcohol is more pleasurable than in the normal population, and this leads to greater alcohol consumption and addiction. This same biological predisposition toward greater pleasure from alcohol is often found in their children, and this may explain (in part) why the children of alcoholics gravitate toward alcohol consumption. Biological and genetic factors do not determine whether a person will be an addict; however, these factors may predispose certain people to addiction. A person's biology affects their predisposition toward addiction; accordingly, their biology can affect their recovery efforts.[83] Treat the body, and consult with addiction doctors to find medications that can assist the addict to reduce the cravings and to address acute withdrawal symptoms (such as anxiety and depression).

Consult with a doctor who has a strong addiction practice and NOT a general doctor. A trained and experienced physician who treats addiction can provide the addict with medicines and treatments that address the biological effects resulting from the cessation of drug use. Addiction physicians have the knowledge and training that the ordinary persons and doctors do not possess, and they have the ability to refer the addict to other persons who can assist the addict (such as counselors, psychologists and psychiatrists). Addiction doctors can address many addiction symptoms, including cravings. In regard to cravings, be aware that an addict's cravings are not the same as the cravings of a non-addict. An addict's cravings are intense desires coupled with a prolonged, continuous obsession. Be alert to the presence of cravings and address the cravings because the addict's intense desire, and their prolonged and continuous obsessions, coupled with the physical effects of withdraw, can fatigue and weaken the addict's conscious efforts to stop their drug use. Addiction physicians may also encounter and treat other brain illnesses. In the same way that you would seek a physician to help you treat your diabetes, you should seek a physician to work with you to treat the addiction.

But note: working with a doctor to treat the addiction does not mean abdicating responsibility to the doctor. Good and great doctors will want to work with you to treat the addiction, and they will seek the addict's input as to the results of the treatment and the medicines. Some of the medicines that are used to treat addiction are also addictive, and therefore it is critical that the doctor possess the specialized knowledge needed to treat addiction.

Biology also extends to other brain conditions and diseases. Biologically, persons with partially-developed minds (teenagers) or mental diseases (including attention deficit disorder or ADD) are at a greater risk of addiction than other individuals. Rather than using drugs to escape the everyday world, the addict may actually be using the drugs to remain a part of and to "cope" with the everyday world.[84] The addict may be using drugs as a form of self-medication to address other brain diseases, even though the drug use may actually exacerbate or trigger other brain diseases. Therefore, to address the existing brain diseases of addiction, it may be helpful to determine whether other brain diseases co-exist with addiction.[85] Consult an addiction physician and asked them to look for the presence of other brain diseases. Some brain diseases may be masked by addiction, and these diseases may not clearly present themselves until the addict has an extended period of sobriety. It is easy to focus on the addiction and to overlook other brain diseases. Therefore, take the time to consult the physician and to ask them to expressly look for the presence of other brain diseases.

Addiction is a brain disease, the brain is a part of your body, and a person's biology affects their addiction; therefore, it's important to address the predisposed biological components that affect addiction and recovery.

The psychological factor refers to the addict's thought processes and emotions.

Initially, the psychological factor centers upon post-acute withdrawal. Post-acute withdrawal follows acute withdrawal, and it occurs because the brain's chemistry is gradually returning to normal. As the brain improves its ability to function, the brain chemicals fluctuate and the fluctuation contributes to post-acute withdrawal symptoms.[86] While the symptoms of acute withdrawal vary greatly between addicts, the symptoms of post-acute withdrawal vary less between addicts. The most common post-acute withdrawal symptoms are anxiety, irritability, tiredness, mood swings, low enthusiasm, variable energy, variable concentration, and the inability to obtain restful sleep.[87]

Post-acute withdrawal symptoms feel like a roller coaster of thoughts and emotions. At the initial stage of post-acute withdrawal, symptoms will change minute-to-minute and then hour-to-hour. However, as the addict is sober for an extended period time, the symptoms will become less severe and may depart or return only sporadically.[88] During the post-acute stage, the periods in which the addict feels "good" and thinks rationally become longer in duration. However, addicts tend to be optimistic, and they need to be on guard against the return of post-acute withdrawal symptoms. There is no single trigger for the return of post-acute withdrawal symptoms, and therefore the addict needs to be vigilant toward the presence of these symptoms. When the symptoms return, the addict needs to know that the symptoms will lessen over time. As the addict begins to experience and to cope with the roller coaster of symptoms, the roller coaster dips will become less severe and the "good" periods will last longer.

Post-acute withdrawal usually lasts from 6 to 18 months. Therefore, it is important for the addict to prepare for an extended period in which their thoughts and emotions will seem to be on a roller coaster. But, since this roller coaster of thoughts and emotions is an indicator that the brain is working to establish a new equilibrium and to recover its previous function, the addict should be made aware that the roller coaster is a sign that the brain is healing.

The addict's psychology will improve as they learn to work and cope with post-acute withdrawal. There are various things that the addict can do to cope with post-acute withdrawal. First, the addict can practice patience. It is difficult for addicts to think in regard to long-term periods of time, therefore the addict should initially practice patience on a short-term / day-by-day basis. As the days pass, the addict will exhibit patience over longer periods of time by building on their day-by-day work. Second, the addict should learn to not fight the symptoms and to instead manage the symptoms. The addict can learn to simply accept the symptoms and to "go with the flow". If the addict consciously learns to "go with the flow", they will learn to minimize the effects of the symptoms. Third, the addict should develop skills that permit them to relax and lessen their stress. The addict needs to understand that becoming rigid and dwelling on the symptoms can make them worse, and that this rigidity can serve as a trigger for relapse. By taking the time to develop behaviors that bring relaxation, the addict can focus on the positive things that will lessen the symptoms and reduce the potential for a trigger that causes a relapse. Fourth, the addict should take some time to celebrate their development of new coping skills. The small celebrations can occur as a trip to the park, a dinner with a friend, or an outing with a group of friends. The addict should acknowledge to themselves that the development of the new coping skills are empowering—for the addict's entire life. These small celebrations can invoke feelings of satisfaction and accomplishment—which improves the addict's psychology.

Another means to improve the addict's psychology is counseling. Counseling can assist the addict to assess their current condition, and to develop strengths that they can use to improve their state of mind. Counselors (which can be doctors, psychologists, psychiatrists, family counselors, or drug counselors) have knowledge that laypeople do not possess, and they can suggest ideas, practices and behaviors that benefit the addict. Since not all counselors are trained to deal with addicts, it is beneficial to perform some research on the counselors in your area who produce good and consistent results. Good counselors will motivate the addict to seek change by using the addict's own motivations and predispositions, and by motivating the addict to seek change to improve their life. The counselor can provide the addict with positive incentives to seek a sustained recovery, and help the addict to develop an understanding of addiction and how the addict can manage their addiction. In this regard, the counselor is a coach, and the coach trains the addict to develop better peace of mind, happiness, self-esteem, satisfaction, the ability to sustain healthy relationships, and the ability to improve their communication. With good counseling, the addict's psychology can be improved, and they can develop the skills and motivation that are required to manage their disease.

In regard to psychology, counselors can develop one especially useful skill—the ability to anticipate the illogical thoughts produced by the brain. The addict's brain has been trained to produce illogical thoughts, and it will take some time for the brain to re-train itself. Accordingly, addicts are often encouraged to disregard the first thought that comes to their mind because the thought is generated by a brain that is affected by addiction. The addict's brain often produces thoughts that perpetuate the addiction and that minimize the addict's perception of the damages caused by the addiction. Despite the fact that the addict's brain can produce thoughts that are illogical, it is predictable that the addict's brain will produce these types of thoughts. Therefore, addicts are often trained to disregard the "stinking thinking" produced by their brain, and to pause and process their second and third thoughts before they take action. As the addict develops this skill, they will gain confidence in their ability to make positive and rational decisions. These positive decisions will lead to results that are better than the results produced by "stinking thinking", and the addict's mental outlook and psychological condition can improve.

Addiction is a brain disease that affects the addict's thoughts and emotions; therefore, it's important to empower the addict to anticipate and manage their thoughts and emotions.

The sociological factor refers to the addict's environment.

The addict's environment and their ability to manage their environment will affect their recovery. The environment includes the people, places, things, employment, family and friends that affect the addict and his behavior. The addict's choice to use drugs was probably influenced by their environment, and therefore the environment must be considered as the addict ceases their drug use. For example, if the addict lived in a family where drug use was common, they need to learn to deal with the family's drug use and to separate themselves from the drug use. Similarly, if the addict had a close circle of friends who all participated in drug use, the addict would need to separate themself from the friends and their drug use. To establish this separation, the addict may need to stay away from certain parts of the city or neighborhood in which they lived, or they may need to simply move away. This sociological change— separation or change in environment—is a means to avoid the triggers that lead to relapse.[89]

For many people, it is not possible to separate themselves from all the people, places, things, employment, family and friends who were a part of their drug use. In these cases, the addict needs to develop the ability to avoid the situations that accompanied their drug use and to set boundaries regarding the environment to which they will expose themselves. For example, if an addict used drugs prior to parties or sports, they should avoid the people, places and things with whom the drug use was associated and they should consciously communicate their new "boundaries" to the people around them. If the addict finds that the people around them cross those boundaries, the addict should reduce the number of instances and the length of time that they spend with those people, or the addict should simply stop associating with those people. To sustain their recovery, the addict needs to change their environment and they should disassociate themselves with the people, places and things that were previously associated with drug use.

To maximize their chance of recovery, the addict should create a new environment for themselves. They should acquire friends who do not use drugs. If possible, the addict should develop friendships with addicts who have several years of sobriety. The addicts who have been sober for many years have the knowledge and skills needed to sustain their recovery, they understand the challenges, and they are generally willing to share their knowledge with people who are new to recovery. In early recovery, it's helpful to live in a sober home because the people who live at the home understand the challenges of addiction and they generally attend the same support group meetings. Further, at sober homes, there are frequent drug tests to ensure that those who live at the home remain drug-free. The drug tests are an element of accountability because people who fail the drug tests must leave the sober home. Creating an environment where drug use is prohibited is helpful to addicts.

To maximize their chance of recovery, the addict in early recovery should focus on their recovery and not on members of the opposite sex. Spouses and other people with whom the addict has an intimate relationship can sometimes provide motivation to the addict. However, addicts who are young (under 25 years of age) or who are early in the recovery are often distracted by intimate relationships. Therefore the addict should focus on their recovery and create an environment that has limited exposure to intimate relationships. If the addict focuses on their recovery, and they develop strong skills and a strong recovery program, the intimate relationships that they thereafter establish will have a better chance of survival. However, intimate relationships that begin too soon will often distract the addict and trigger a relapse.

Addiction is a brain disease that is affected by the addict's biology, thoughts and emotions, and their environment; and, training the addict to manage their environment improves the addict's ability to manage and sustain their recovery.

To improve the chances of successfully using the stages of recovery, consider the three primary factors that affect addiction. Recovery requires a balanced approach to the biological, psychological and sociological factors in the addict's life. These three factors are very important in the treatment of addiction.[90] No factor should be ignored or over-weighted in its importance. All three factors should be used in a balanced way to support the addict's recovery program. Utilize the three factors and encourage the addict to tailor them to his or her specific needs.

Seeking Recovery: Some Suggestions

Given the complexity of the disease, this book will <u>not</u> provide you with a recommended course of action regarding your specific recovery. The right course of conduct will vary based upon many factors, including the addict's age, sex, socio-economic status, drug of choice, stage of recovery, your family history, and many other factors. Nevertheless, we will provide you with some suggestions that you may wish to investigate and consider. In that regard, our first suggestion is that you become familiar with the disease of addiction, and that you obtain assistance from people who are familiar with the disease of addiction. The addict may begin by simply attending Alcoholics Anonymous or Narcotics Anonymous meetings, and you may begin by attending Al-Anon meetings. You (and the addict) may also choose to seek assistance and advice from a physician or drug counselor. You (and the addict) may consult various resources using the Internet, and gain additional insight into the disease. In all cases—you should become familiar with the disease of addiction.

Our second suggestion is that you seek help. Some addicts and their families choose to fight addiction without professional help, and some succeed. But, the clear facts are that most addicts succumb to their disease, the thoughts produced by the addict's brain perpetuate the disease, the scientific knowledge regarding addiction is still developing, and there is no cure for addiction. So, if you want to improve the addict's and your changes of recovery, we recommend that the addict and their family seek professional assistance.

In regard to the second suggestion, many addicts begin their treatment in the office of their family physician. Addiction is an illness, and therefore many addicts choose to obtain treatment from their family physicians. If the addict visits with their family physician for treatment, consider asking for a referral to an addiction specialist. You (or the doctor) may discount the need for a referral. But, consider the following: if the family physician encountered a strong psychological, neural or other brain disease, they would probably make a referral to a specialist. The situation is no different here because addiction is a <u>brain disease.</u> Physicians with an addiction specialty possess knowledge and experience regarding addiction that other physicians do not possess, and the addiction specialists have strong knowledge of the science of addiction, addiction programs, support groups, and other addiction professionals who can assist the addict. Rather than trying to acquire this specialized knowledge while treating the addiction, consider availing yourself of the resources in your community that already possess the specialized knowledge and the contacts that can improve your treatment program.

The third suggestion is that you consciously consider that you are seeking treatment for the disease of addiction, and that your "help" to the addict should ***<u>exclude</u>*** solving the addict's problems or shielding the addict from the consequences of their actions. In the case of Diabetes—especially if the person was an adolescent—you would seek a treatment that will improve the condition of the disease; you would try to lessen the burden for the patient "dealing" with the disease; and you would try to protect the patient. This process does <u>not</u> work with addiction. You need to ensure that the addict is the person in charge of their recovery. You can be the person in charge of your recovery—but the addict must be in charge of their recovery. You need to help the addict recognize the existence of the disease, and to permit them to make the decision to seek recovery treatment based upon <u>their</u> motivations, desires and needs. You can help the addict see the need for recovery; but, ONLY the addict can choose to become sober and stay sober. Similarly, <u>only</u> YOU can choose to seek recovery and a more normal life for YOURSELF. When dealing with the disease of addiction, each person must choose whether they will seek recovery. If the addict is your teenage child, following this suggestion is especially difficult—seek professional help.

Recovery Treatment Programs

For many people, it is helpful to obtain assistance from an inpatient or outpatient treatment program. These programs vary widely, and considerable information is on the Internet regarding treatment programs. Generally, inpatient or outpatient treatment programs guide the addict through detoxification and acute withdrawal, and then provide the addict with recovery skills and knowledge to manage post-acute withdrawal.

As you review the treatment programs, begin by considering the detoxification process. There are a number of medicines that can lessen the ill effects of detoxification and assist the addict through acute withdrawal—but whether the addict needs these medicines depends upon the addict. Also, <u>be aware that withdrawal for alcohol, barbiturates and other depressants can cause death;</u>[91] therefore, <u>seek professional help when dealing with persons who are addicted to alcohol, barbiturates or other depressants</u>.

When assessing an inpatient or outpatient program, the methods, practices and policies of the programs vary greatly. Therefore, consider at least the following factors:

1. The reputation of the program. Review the reputation of the program with drug recovery organizations and the physicians who practice in the area of addiction.
2. The impressions of persons who attended and graduated from the program. Addicts are very good sources of information, and they can possess strong feelings and impressions of different programs. Consider locating and speaking with graduates of the treatment programs, and becoming familiar with their viewpoints regarding the effectiveness of the program.
3. The demographics of the persons who attend the program. Review and consider the program's experiences with the sex, age and medical condition of your addict.
4. Use of medication. Assess the views of the program regarding the use of medication. Some programs prohibit the use of any medications that produce mind-altering effects, and other programs permit medications (even if they produce mind altering effects) that are a part of a drug withdrawal or harm mitigation program. Neither option is without risk; therefore, we recommend that you discuss both options with an addiction doctor.

5. The daily schedule. Review the daily routine of the program, and the specific results that the program seeks to achieve. In this regard, you may want to become familiar with the after-care provided by the program and its alumni networks.

6. The recovery process. Ask the program to describe their program, from admission to release. Ask the program to describe the "typical" parts of each day, and how these parts will change each week. Ask the program to describe how the program's process addresses and treats the illness of addiction, and how the addict's progress in the program will be measured and how their treatment will be adjusted based upon the measurements. Assess the program's process, and how the process meets the needs of the addict.

7. The program's deliverables. In order for the program to affect the addict in the desired way, the program must set the desired outputs or deliverables—those things that the program will instill in the addict and empower them to manage their addiction. Ask the program to describe its goal (probably sobriety and the means to maintain sobriety) and desired outputs or deliverables that will lead to the goals. A program with an ill-defined goal and ill-defined outputs will have a reduced chance of "success" because (a) it failed to set the measures by which success toward the goals will be assessed and (b) the program is not designed to achieve the desired goals or outputs/deliverables.

8. Family care. Many programs also include a family treatment program. Addiction is a disease that affects the family, and the family generally needs treatment. It is helpful to assess how the addict's recovery program for the addict overlaps with the family recovery program, and how the intended results of each program complement each other.

9. The program's credentials and reputation. Review the credentials of the program with state and local agencies, and review the reputation of the program with the local chapters of national drug recovery organizations.

The factors described above will help you to begin your assessment; however, the assessment must consider the specific needs of the addict and the family.

The common length of an inpatient treatment program is 30 days; however, a 90-day inpatient program can provide many benefits over a 30 day program. In that regard, consider that during the first two weeks of treatment, the addict is typically in acute withdrawal and the addict is "in a fog". The symptoms of acute withdrawal limit the ability of the addict to understand the information that they are being provided by the treatment program. Every addict is different, but many addicts do not begin processing the information and lessons that they are being provided until the third or fourth week of treatment. This means that although the addict has undergone detoxification and spent 30 days in treatment, they are being released from a 30 day program with just 14 days of treatment that the addict can recall. We suggest that you consider a 90 day program, and that the addict "graduate" from the 90 day program to a sober living home. We suggest that the addict take some time to permit his or her brain to heal, and to develop a recovery program based upon a new sober-living lifestyle. This process is not ideal for all addicts—and it can be expensive—but it also serves to detoxify the addict, to provide him or her with some basic information and sobriety lessons, and to place them in a sober living environment where the common routine involves the addict's use of the information and lessons that they learned while in the recovery program.

There is no single means to assess the suitability of a program. We recommend that you collect and assess the factors that are set forth in the preceding paragraphs. Further, consider condensing the review to some key items by telling the program to set "sobriety and the means to maintain sobriety" as the goal, and to describe: (1) What knowledge, skills, behaviors will the program designed produce in order to satisfy the goal? (2) How will the program produce these behaviors and outcomes? (3) How will the addict use these behaviors and outcomes? (4) How will the program and the addict deal with a relapse? Assess the specificity and the completeness of the responses, and how the responses address your needs.

IMPORTANT: Please beware that, when assessing different programs, setting the goal of "sobriety and the means to maintain sobriety" does NOT mean that the program can guarantee sobriety. The program can produce knowledge, skills, lessons, and behaviors that the addict can use to obtain and maintain sobriety, but only the addict can choose to obtain and maintain sobriety.

Recovery: The Importance of a Good Maintenance Program

An important key to sustained recovery is a good maintenance program.[92] After the addict exits a treatment program, they often continue with a 12 step program, and addicts that do not participate in a treatment program also often begin their sobriety with a 12 step program—they end up at the same place. A 12 step program is a self-help support group based upon the 12 steps and the ideology of Alcoholics Anonymous (AA). From its website, Alcoholics Anonymous states that it is a fellowship of men and women who share their experience, strength and hope with each other so they may solve their common problem and help others to recover. The website states that the only requirement for membership is a desire to stop drinking (or using drugs). At an Alcoholics Anonymous meeting, those in attendance will proceed through a structured agenda. The typical meeting begins with the reading of the AA preamble and the AA 12 steps. A meeting will often include the celebration of different sobriety anniversaries. There are typically one or two speakers who will share their experience with the group, and newcomers are encouraged to share how many days that they have been sober. Meetings typically begin and end with a group prayer and the recitation of AA slogans.

Alcoholics Anonymous meetings focus upon a 12 step program.[93] Alcoholics Anonymous and the 12 steps were developed by two former alcoholics, who discovered that they were able to maintain sobriety more frequently and for longer periods of time by working together (as compared to trying to maintain the recovery alone). At the meetings, speakers will share their experiences and challenges, and attendees can (a) gain insights from these experiences and challenges and (b) selectively apply the insights to their own situation. In the meeting, and after the meeting, addicts can share their experience working the 12 step program, and they can gain insights and lessons from the experience of others. A copy of the 12 steps is set forth as Exhibit E.[94] The 12 steps and AA include references to God, and the 12 steps and AA include spiritual and religious elements. The spiritual and religious elements may not be appropriate for everyone; but, when addiction has filled your life with hopelessness, the spiritual and religious elements can often be used to regain some hope and serenity. The information obtained at

Alcoholics Anonymous meetings can be anecdotal; but, over time, these meetings provide the addict with useful information regarding addiction and how to manage it.

EXHIBIT E: THE AA 12 STEPS

THE TWELVE STEPS OF ALCOHOLICS ANONYMOUS

1. We admitted we were powerless over alcohol—that our lives had become unmanageable.

2. Came to believe that a Power greater than ourselves could restore us to sanity.

3. Made a decision to turn our will and our lives over to the care of God *as we understood Him.*

4. Made a searching and fearless moral inventory of ourselves.

5. Admitted to God, to ourselves, and to another human being the exact nature of our wrongs.

6. Were entirely ready to have God remove all these defects of character.

7. Humbly asked Him to remove our shortcomings.

8. Made a list of all persons we had harmed, and became willing to make amends to them all.

9. Made direct amends to such people wherever possible, except when to do so would injure them or others.

10. Continued to take personal inventory and when we were wrong promptly admitted it.

11. Sought through prayer and meditation to improve our conscious contact with God, *as we understood Him,* praying only for knowledge of His will for us and the power to carry that out.

12. Having had a spiritual awakening as the result of these Steps, we tried to carry this message to alcoholics, and to practice these principles in all our affairs.

Copyright © A.A. World Services, Inc.

AA meetings are also places at which attendees can become familiar with various slogans. The slogans are geared toward maintaining sobriety, and they encapsulate certain lessons that can be readily recalled and applied if the addict is experiencing problems maintaining their sobriety. For example, one slogan is "Think it through", which reminds the addict to avoid the first thought that comes to mind because the mind may romanticize the drug use experience. "Think it through" reminds the addict of the negative consequences that result from acting upon the first thought that comes to mind.[95] Another slogan, "Stinking thinking" reminds the addict that their first thought may not be a useful thought because it is generated by a brain that is affected by addiction. By attending Alcoholics Anonymous meetings, addicts often learn to discount the first thought, and to think past the first thought and through to the consequence of the actions.

Alcoholics Anonymous has been in existence since 1935, and it has meetings throughout the world.[96] Because Alcoholics Anonymous meetings are present in many places and at many different times of the day, and because they are conducted in the same general manner, this is a self-help program that addicts can readily locate and attend regardless of where they live. Moreover, since Alcoholics Anonymous includes special meetings that are geared toward women, men, teenagers, mature adults, and other subgroups, the addict can readily become part of a group with which they can possess some affinity. Alcoholics Anonymous provides the addict with a new group of friends, and Alcoholics Anonymous provides a way for the addict to replace the old group of friends that engaged in drug use with a new group of friends that support sobriety. Alcoholics Anonymous, and the fellowship of the people who attend its meetings, is available throughout the world.

Alcoholics Anonymous meetings focus on the experiences and practices that have helped addicts obtain and sustain sobriety, as compared to the science of addiction. Alcoholics Anonymous meetings do not focus on neurons and neurotransmitters. Nevertheless, the experiences and practices shared at Alcoholics Anonymous meetings, and the additional information that can be obtained from the organization and its website, have been revised and refined since 1935. Therefore, Alcoholics Anonymous can provide many benefits to you.

Our advice is that you make use of all reliable resources that help addicts to obtain and sustain sobriety. Therefore, in addition to AA, there are other self-help groups to support recovery. Those self-help groups include Secular Organizations for Sobriety (SOS), SMART, and Women for Sobriety.

SOS is an alternative 12 step program which omits the spiritual and religious aspects of Alcoholics Anonymous.[97] SOS supports the premise of Alcoholics Anonymous that life-long abstinence is necessary to maintain sobriety and that a support group is very helpful. But, unlike Alcoholics Anonymous, SOS believes that the 12 steps are not appropriate for everyone, and SOS encourages its members to try different methods to maintain their sobriety until they find the specific approach that works for them. SOS supports the idea that alcoholism and drug addiction is fundamentally a health problem, and they therefore discount the need for spiritual or religious components as a part of the recovery program. Instead, SOS supports behavioral modification by re-training of the "higher" cortex area of the brain (i.e. the neo-cortex) to restrain the "primitive" lower part of the brain (i.e. the limbic system).

SMART is based upon the principles of cognitive behavioral and motivational enhancement therapy.[98] SMART teaches that no "higher powers" are required for recovery, and that addicts must focus on their own competence to recover. SMART does not have "steps"; instead, they focus on the following four areas to develop and maintain sobriety: (1) build and maintain the motivation needed to gain independence from addiction; (2) cope with urges; (3) manage thoughts, feelings and behaviors; and (4) live a balanced life. SMART seeks abstinence, but it does not completely reject moderate drug use. Also, SMART meetings focus on self-empowerment, education, and the managed use of medicines.

Women For Sobriety recognized that many of the principles of Alcoholics Anonymous did not meet the needs of women, and therefore this program is tailored to address the unique needs of women addicts.[99] Women For Sobriety focuses on the capability, competence, caring and compassion of the women addicts, and it minimizes the need to focus on humility and flaws (since women commonly over-focus on these attributes). Women For Sobriety also place additional emphasis on meditation, diet, exercise and positive thinking exercises. Women For Sobriety has 13 affirmations (called the new life acceptance program) rather than the 12 steps from Alcoholics Anonymous. [100]

Another set of self-help groups come from religious organizations. A large number of churches, synagogues and mosques include programs to assist their membership in drug recovery. If the addict or the family of the addict have a strong religious affiliation, it may be useful to consult with their religious leader regarding the recovery programs offered by the religious organization.

Harm Reduction

Up until this point we have spoken solely about the disease and recovery. There is another option that you should understand: harm reduction. Harm reduction seeks to minimize the harm related to drug use, and it seeks to connect the addicts with services that supplement abstinence-oriented treatment.[101] Harm reduction recognizes that some addicts may never be able to abstain from drug use or they may take an extensive period of time before they can abstain, and it therefore seeks to minimize the harm caused by drug use. In addition to minimizing the harm, harm reduction seeks to keep the addict alive for a sufficiently long period of time to provide them with more opportunity to seek recovery. The concept is to provide an offer of hope to the addict. However, this practice can result in enabling behavior, and therefore extreme care should be exercised if you consider the use of harm reduction.

Harm reduction seeks to minimize the damage caused by drug use, and it provides various practices that can minimize the damage. For example, snorting drugs is usually less harmful than injecting the drugs, and therefore the drug addict could be advised to avoid injecting drugs when they can be snorted. If the addict uses needles to inject the drugs, they could be advised to clean the skin at the injection site before injecting the drugs. Addicts who use needles could also be advised to clean the needles with bleach and to thoroughly rinse them with water before they are reused. Further, addicts that use needles could be advised to not share needles with other addicts, and to exchange the needles at facilities that provide needle exchanges. In a more positive manner, active drug users could be encouraged to switch to less harmful drugs or to use drugs less frequently. In order to increase the periods between drug use, the addict can be advised to seek more work or more play in order to distract their mind from drug use. All of these options are not directly aimed at recovery—their focus is on merely minimizing the harm caused by drug use.

It is important to note that harm reduction can conflict with the traditional teachings and practices of 12 step programs. The steps (and the 12 step programs) emphasize that addicts who are using drugs are powerless over the drugs and out of control, but harm reduction pre-supposes that addicts can make <u>some</u> choices to affect their drug use. Harm reduction is a relatively new concept, and it is <u>readily</u> susceptible to abuse by addicts in order to control their families and friends. Therefore, although you should be aware that the practice of harm reduction exists, you should seek professional assistance when considering the use of harm reduction. In our view harm reduction is very susceptible to abuse, and it should be used cautiously.

Use of Drugs to Treat Drug Addiction

An important subject matter that you should understand is the use of drugs to treat drug addiction. There are various drugs that physicians can use to address the biological effects of drug use and to reduce cravings, and to make the doctor's treatment of the psychological and sociological factors more effective. For example, for opiate addicts, Methadone, Bupronehine (also known as Suboxone) and Naltrexone are opiate blockers that block some of the "high" produced by opiates and lessens cravings.[102] Methadone and Bupronehine are based upon opiates and produce some mind-altering feelings, and therefore they can become addictive too.[103] Naltrexone is not based upon opiates, but it does not prevent addicts from switching to other drugs.[104] There is a new version of Naltrexone called Vivitrol that blocks the effects of opiate for one month.[105] Also, for alcohol addicts, physicians can use Naltrexone (described above) or Disulfirum (also known as Antabuse).[106] Disulfirum causes the addict to become physically ill if they consume alcohol. Remember, there is no single treatment that is appropriate for every addict, and there is no magic pill or treatment program that is a cure. Addiction is a complex but treatable disease; and, to increase the addict's chances of a sustained recovery, you should consider the use of doctors who can carefully utilize medicines and procedures to treat addiction.

Family Recovery and Al-Anon

A very useful program for the families of addicts is Al-Anon. Al-Anon is a mutual support group for persons who experienced problems with someone else's addiction.[107] Al-Anon focuses the recovery of the family members rather than the addict. Indeed, Al-Anon members will discover information and practices that can assist them to live a more normal life even if the addict never seeks recovery. Meetings are held in different geographic areas within a city or county, and some meetings specifically focus on problems experienced by parents, spouses or children. There is also a group called Ala-teen for teenagers whose parents or friends are afflicted by addiction. Each meeting has the autonomy to be run as its members choose within the Al-Anon guidelines, and therefore the meeting structure and content will vary. Al-Anon meetings often begin with a reading of the Twelve Steps of Al-Anon, the Twelve Traditions, and the Alcoholic's Letter (set forth below).[108] Members do not provide advice, but they share their experiences and lessons with other members. Members are encouraged to adapt and use the experiences and lessons that they find to be useful and to leave the rest. In addition, Al-Anon has many useful pamphlets, books and audio recordings for people to use. Al-Anon is a good mutual support group with support and informational materials.

In Al-Anon, members do not provide advice; however, there is one Al-Anon document that does provide specific guidance: the Alcoholic's Letter. This document (which generally applies to addicts) provides specific insights into the thoughts and behaviors of an addict, and the document describes how the people around the addict should behave when interacting with the addict. The Alcoholic's Letter reads as follows:

AN OPEN LETTER FROM AN ALCOHOLIC

I am alcoholic. I need your help.

Don't lecture, blame or scold me. You wouldn't be angry at me for having cancer or diabetes. Alcoholism is a disease too.

Don't pour out my liquor; it's just a waste because I can always find ways of getting more.

Don't let me provoke your anger. If you attack me verbally or physically, you will only confirm my bad opinion about myself. I hate myself enough already.

Don't let your love and anxiety for me lead you into doing what I ought to do for myself. If you assume my responsibilities you make my failure to assume them permanent. My sense of guilt will be increased, and you will feel resentful.

Don't accept my promises. I'll promise anything to get off the hook. But the nature of my illness prevents me from keeping my promises, even though I mean them at the time.

Don't make empty threats. Once you have made a decision, stick to it.

Don't believe everything I tell you; it may be a lie. Denial of reality is a symptom of my illness. Moreover, I'm likely to lose respect for those I can fool too easily.

Don't let me take advantage of you or exploit you in any way. Love cannot exist for long without the dimension of justice.

Don't cover up for me or try in any way to spare me the consequences of my drinking. Don't lie for me, pay my

53

bills, or meet my obligations. It may avert or reduce the very crises that would prompt me to seek help. I can continue to deny that I have a drinking problem as long as you provide an automatic escape for the consequence of my drinking.

Above all, do learn all you can about alcoholism and your role in relation to me. Go to open AA meetings when you can. Attend Al-Anon meetings regularly, read the literature and keep in touch with Al-Anon members. They're the people who can help you see the whole situation clearly.

I love you.
Your Alcoholic

The Alcoholic's Letter provides specific insights and guidance. We recommend that you make a copy for yourself, and that you consult it from time to time.

Below is our interpretation of the Alcoholic's Letter (which applies to drug use as well). Al-Anon maintains that members should "take what they can use, and leave the rest"; accordingly, in that spirit, take from the following interpretation of the Alcoholic's Letter what you can use and leave the rest. The Alcoholic's Letter guides the family member in what they should do and what they should NOT do. The first lesson provided by the Alcoholic's Letter is that addiction is a disease, and that it should be treated like a disease. Supply of the drug is not the problem—the brain disease is the problem. The second lesson is that family members should avoid enabling the addict. The addict will make statements, provide arguments, and engage in behaviors that promote enabling behavior by family members. The Alcoholic's Letter provides clear examples of enabling behaviors that should be avoided. As a family member, focus on what you can see and verify—not the addict's statements or promises. The third lesson is that family members need to set the boundaries required for the family to live a more normal life, and they should enforce those boundaries no matter the consequence to the addict. Setting and enforcing boundaries is helpful to families; and, addicts who cannot readily manipulate their families into enabling behaviors will have fewer options—and the addicts may therefore seek recovery. The fourth lesson is that only the addict can obtain and maintain their sobriety, and that the addict must therefore (a) find their own motivations to obtain and maintain their sobriety and (b) experience the consequences of failing to obtain and maintain sobriety. Addicts can be motivated by Hope and Fear (which are primary emotions)—but NOT usually by shame or family anger. Permit the addict the opportunity to find the motivation that leads them to obtain and maintain their sobriety. The fifth lesson is that family members should seek out information about addiction, use that information to develop a recovery program for the family, and create an environment where the addict is more likely to obtain and maintain sobriety. Take from this interpretation of the Alcoholic's Letter what you can use and leave the rest.

Our goal is to inform you why it is very difficult for the addict to stop, and to provide you with the answer in a short book. We are tempted to provide more specific guidance as to how to seek family recovery and how to manage your relationship with the addict. However, in order to fulfill our primary goal and to keep the book as short as possible, we elected to not provide more specific guidance. Instead, we recommend that you seek professional addiction treatment, you read the books that are footnoted in this book, and you consider the general guidance provided by this book.

Conclusions

This book has attempted to provide a basic understanding of addiction, and provide some suggestions that you may find to be helpful. However, as in Al-Anon, this book does not seek to provide specific recommendations or guidance that will apply to all addicts. Addiction is a complex disease, and it needs to be treated on an individualized basis. Further, our advice is that you also seek assistance and guidance from professionals who treat addiction.

We offer, for your consideration, a four-step process that we used to create and maintain our family recovery program. We believe that this four-step process has applicability to the family <u>and</u> the addict. The steps are: (1) Convert thoughts into thinking; (2) Convert thinking into purposeful action; (3) Convert purposeful action into skills and tools; and (4) Use your skills and tools to build and maintain your recovery program and your new way of life. In addiction, people instinctively react to urges and cravings, and they need to learn to slow down and engage the neo-cortex in conscious thinking. The conscious thinking can help to suppress the unconscious actions driven by the limbic system, and produce purposeful action. As the brain is repeatedly engaged in thinking and purposeful action, the addict (and the addicted family) develops and practices their skills and tools. And, as the skills and tools are successfully and repeatedly used, the addict and the addicted family can build and maintain a recovery program and a new way of life. These steps require knowledge of addiction, including various skills and tools; therefore, if you use these four steps, we suggest that you educate yourself regarding addiction and that you work with addiction professionals. Work with addiction professionals because the knowledge that they possess can help you, and the knowledge that you do not possess can hurt you.

Addiction affects the entire family. You have no power over the addict—they must choose to seek recovery or not. But, your life can get better—you have the power to choose recovery for yourself. Seek recovery for yourself.

One of the hardest things to understand is that to seek recovery, you must have hope and you must set aside your pre-established dreams for the addict. Set aside all of the education, trips, experiences, jobs, and relationships that you dreamed would enrich the life of your addict <u>before</u> the addiction. Many people obsess over these dreams, and they try to preserve the pre-addiction dreams by "helping" the addict. Unfortunately, this helping often manifests itself as enabling.

Helping an addict means that you provide the addict with choices, and you allow them to choose and to experience the consequences of their choices. Enabling is making the choice for the addict, and allowing the addict to avoid the consequences of their choices. But note: the distinction between helping and enabling is clearer and easier to discern when the situation is abstract or it pertains to someone else—the distinction is much harder when it concerns your addict, it affects the addict's and your life, and you have only a few seconds to make the choice.

Set boundaries to pre-establish choices, and practice making choices. These choices are a part of your recovery—so reflect on your choices and correct them if required. Practice making more informed choices, and do not fear correcting mistaken choices. You deserve hope and a better life—make choices that preserve and grow hope for you and your dreams for your life.

In some cases—many cases—when you make more informed choices and you work your recovery program, the addict learns that their options are fewer and the consequences of their drug use are more immediate and more direct. When you make more informed choices and work your recovery program, then sometimes the addict will find the need and the motivation to seek recovery for themselves.

Addiction is a life-long disease, and it has no cure.[109] Nevertheless, the persons who aggressively follow a recovery program have a significantly better probability of living a better life. Therefore, improve the odds—aggressively follow your recovery program, and encourage the addict to aggressively follow their recovery program.

If you learn to make informed choices, you make those choices, and you do your best—you have done all that is possible. Accept that you are powerless over the disease and the choices made by the addict, but find peace in the fact that you did all that was possible and you did your best.

There are many good sources of information regarding addiction, and two sources are especially informative. Use the Internet to access the websites of (a) The Substance Abuse and Mental Health Services Administration (www.samhsa.gov) and (b) the National Institute on Drug Abuse (www.drugbuse.gov). The organizations and their website provide accurate, informative and up-to-date information, and they also have publications that you can obtain, read and use.

This book is a work-in-process. New advances in addiction treatment are made, and we will endeavor to keep abreast of those advances. We will endeavor to revise this book, to post new information, and to cover new topics. Aggressively follow your recovery program; find and grow hope for yourself; and create and fulfill new dreams that will enrich your life. Be an addiction survivor—manage the disease.

Research into addiction is also a work-in-process. This work is greatly affected by the priorities set by the legitimate drug industry, medical universities, medical insurance companies, and the government. Accordingly, even though there is no cure for addiction, call, write or email your elected federal and state representatives and pose the following questions to them:

> About 10% of all Americans have an alcohol or drug addiction. If we assume that the addict lives within a family of four, that means that about 40% of everyone in the United States—and about 40% of your constituents—are affected by addiction. **40%!**

> We spend billions of dollars on the war on drugs to stop the planting, harvesting, distribution and sale of drugs. We spend billions of dollars to reduce the **supply** of drugs and the problem is getting worse. Are you willing to spend an **equal** amount to find a cure for addiction and to dramatically reduce the **demand** for drugs? If no, why not? If yes, when?

> P.S. If a cure drops the **demand** for drugs, then the prices AND the **supply** of drugs will drop. So, since we are on the verge of finding a "cure" for AIDS—why not a cure or effective treatment program for addiction?

Addiction touched your life—so take the action required to manage the effect of addiction upon your life and to drive the change required to find a cure for it (or at least a means to reliably treat more people). You can manage your recovery, and you can also drive the change required for the world to understand that addiction is a disease that needs effective treatment.

Final Recap

You probably read this book in a short period of time; but, it may take some time to understand the information and to effectively apply the information. So, to simplify and highlight specific information, let's summarize the content of the book.

- Why can't the addict stop? The simple answer is that:

    o   drug use changed the addict's brain and
    o   the brain's normal thought processes are changed to produce thoughts that encourage drug use.

- Focus on the root cause of the problem—the brain. Treat the brain and the body—and avoid focusing on the symptoms.
- The book seeks to provide you with information and suggestions that <u>you</u> can use to guide <u>your</u> decisions and actions, and to empower <u>you</u> to obtain some normalcy in <u>your</u> life. You cannot force, cajole or entice the addict to seek and maintain their recovery—that power rests <u>solely</u> with the addict.
- Drugs alter the physical condition of the brain AND the manner in which it produces thoughts. These changes to the brain affect the addict's thoughts, emotions, perceptions, predispositions, and actions. Since the addict's brain and thoughts are changed, the addict is directed by their brain to take actions that are perceived as useful and rational by the addict (and harmful and irrational by us).
- The basic areas of the brain that are affected by drug abuse are the brain stem, the limbic system, and the cerebral cortex.

    o   The brain is structured to learn from positive and negative experiences, and to repeat those activities that are "perceived" as positive and that produce pleasure, reward and continued life. The brain records experiences, it remembers those experiences, and it teaches us to repeat the experiences and behaviors that it perceives are positive. Drugs utilize this same process to teach the brain to produce thoughts which perceive drug use as a positive behavior.

o   The addict's brain will produce thoughts that we see as illogical and irrational. But, to the addict, the thoughts are logical and rational, and the thoughts support continued drug use. Your logic—no matter how profound or compelling—will not cause the addict to become and stay sober. The addict's brain—not your brain—must find the rationale and the motivation to seek recovery and to stay sober.

o   The addict's brain produces thoughts that lead the addict to take actions that the addict's brain perceives are correct, but which we perceive as evidence of a lack of self-control and poor decision-making. The addict's brain is taught to process thoughts that link drug use with the solving the addict's challenges and problems *rather than* being the source of the addict's challenges and problems.

o   The brain learns to instinctively associate the drug use with pleasure and as a critical part of everyday life, and the brain is taught to create lessons and thoughts that reward drug use. In short, the brain changes and learns to produce thoughts that directly link drug use with pleasure and survival.

- **Addiction is a brain disease,** and it produces real and profound changes in the brain and the thoughts produced by the brain. Addiction is NOT a moral affliction. Addiction is NOT an indicator of intelligence. Addiction is NOT a criminal offense. Addiction is brain disease, and the disease should be treated as a disease. Treat the disease, and the symptoms can improve.
- The decision to use drugs was a conscious choice. But, there is no conscious choice by the drug user to become a drug addict. The change from drug user to drug addict is made by the body—it is not a choice. Addiction is NOT a choice.
- There are (at least) five barriers to recovery. The addict's choices are clouded by at least five factors: (1) withdrawal, (2) age, (3) other brain illnesses, (4) friends, and (5) family. These factors

should not prevent the addict and their family from taking action; however, these factors should be considered because they will affect the addict's thoughts, perceptions, and willingness to seek recovery.

- Recovery generally happens in six stages. The stages are pre-contemplation; contemplation; preparation; action; maintenance; and relapse. The addict seeking recovery may spend considerable time in one stage, and less time in another stage. Also, the addict may partially proceed to a higher stage and become stuck between two stages. Further, the addict may successfully proceed toward a higher stage, and then fall back one or more stages. Recovery rarely occurs in an orderly, serial and straight-line basis. Hope for progress; prepare for setbacks; and seek ways to motivate the addict (and yourself) to keep focused on seeking recovery for their (and your) own reasons.

- Recovery requires a balanced approach to the biological, psychological and sociological factors in the addict's life. These three factors are very important in the treatment of addiction. No factor should be ignored or over-weighted in its importance. All three factors should be used in a balanced way to support the addict's recovery program. Utilize the three factors and tailor their use to the specific needs of the addict.

- Given the complexity of the disease, this book will <u>not</u> provide you with a recommended course of action regarding your specific recovery. The right course of conduct will vary based upon many factors, including the addict's age, sex, socio-economic status, drug of choice, stage of recovery, your family history, and many other factors. Nevertheless, we will provide you with some suggestions that you may wish to investigate and consider.

  o Become familiar with the disease of addiction, and obtain assistance from people who are familiar with the disease of addiction. The addict may begin by simply attending Alcoholics Anonymous or Narcotics Anonymous meetings, and you may begin by attending Al-Anon meetings. The addict (or you) may also choose to seek assistance and advice from a physician or drug counselor. The addict (or you) may consult various resources using the Internet, and gain additional insight into the disease. In all cases—you should become familiar with the disease of addiction.

o Seek help. Some addicts and their families choose to fight addiction without professional help, and some succeed. But, the clear fact is that most addicts succumb to their disease, the thoughts produced by the addict's brain perpetuate the disease, the scientific knowledge regarding addiction is still developing, and there is no cure for addiction. Rather than trying to acquire specialized knowledge regarding addiction before proceeding, consider availing yourself of the resources in your community that already possess the specialized knowledge and contacts that can improve your treatment program. We recommend that the addict and their family seek professional assistance, and that they become familiar with the self-help groups that assist addicts and their family (including Alcoholics Anonymous, Secular Organizations for Sobriety (SOS), SMART, and Women for Sobriety, and Al-Anon).

o Consciously consider that you are seeking treatment for the disease of addiction, and that your "help" to the addict should exclude solving the addict's problems or shielding the addict from the consequences of their actions. You can help the addict see the need for recovery; but, ONLY the addict can choose to become sober and stay sober. Similarly, ONLY you can choose to seek recovery and a more normal life for yourself. When dealing with the disease of addiction, each person must choose whether they will seek recovery.

- Addiction affects the entire family. You have no power over the addict—they must choose to seek recovery or not. But, your life can get better—you have the power to choose recovery for yourself. Seek recovery for yourself.
- One of the hardest things to understand is that to seek recovery, you must have hope and you must set aside your pre-established

dreams for the addict. Set aside all of the education, trips, experiences, jobs, and relationships that you dreamed would enrich the life of your addict <u>before</u> the addiction. Many people obsess over these dreams, and they try to preserve the pre-addiction dreams by "helping" the addict. Unfortunately, this helping often manifests itself as enabling.

- Addiction is a life-long disease, and it has no cure. Nevertheless, the persons who aggressively follow a recovery program have a significantly better probability of living a better life. Therefore, improve the odds—aggressively follow your recovery program, and encourage the addict to aggressively follow their recovery program.

- If you learn to make informed choices, you make those choices, and you do your best—you have done all that is possible. Accept that you are powerless over the disease and the choices made by the addict, but find peace in the fact that you did all that was possible and you did your best.

- There are many good sources of information regarding addiction, and two sources are especially informative. Use the Internet to access the websites of (a) The Substance Abuse and Mental Health Services Administration (www.samhsa.gov) and (b) the National Institute on Drug Abuse (www.drugbuse.gov).

- Call, write or email your elected federal and state representatives and encourage them to fund research regarding addiction.

- Drug abuse can make you feel helpless and alone. However, you are <u>not</u> helpless and you are definitely <u>not</u> alone.

- Addiction touched your life—so take the action required to manage the effect of addiction upon your life and to drive the change required to find a cure for it (or at least a means to reliably treat more people). You can manage your recovery, and you can also drive the change required for the world to understand that addiction is a disease that needs effective treatment.

You do not need to memorize all of the facts and lessons described in this book. It is more important to <u>remember, use and practice</u> those facts and lessons that are helpful to you. One of the reasons that addicts are strongly encouraged to continue attending support meetings (such

AA) is for them to continue to remember, use and practice the facts and lessons that will enable them to maintain their sobriety. This same is true for the families of addicts. So, <u>remember, use and practice</u> those facts and lessons that are helpful to you.

Drug abuse can make you feel helpless and alone. However, as you have read and discovered, you are <u>not</u> helpless and you are definitely <u>not</u> alone. We hope that the information in this book provides you with basic knowledge that you can use—so that you can address the disease of addiction and not its symptoms. We hope that you gain Hope that your life can get better. We also hope that you find the Courage to press forward, acquire more information, develop a strong recovery program, and drive change in your family and our society regarding addiction. You can do it . . . step forward one day at a time.

This book is a starting point for recovery. So, become educated; drive change; and be hopeful and brave.

Appendix

The brain performs its functions by collecting and transmitting messages throughout the entire body though chemical and electrical processes, and by changing the brain so that it can guide future actions.[110] The brain's ability to perform these functions is dependent on billions of nerve cells called neurons.[111]

Neurons collect, transmit and process messages, and it relies upon dendrites, axons, terminals and the cell body to perform these tasks.[112] Neurons have a forest-like area at the end of the nerve cell called dendrites, and dendrites receive messages. The message is carried to the cell body along the axon and the neuron's cell body consolidates and processes the messages. The messages are then carried to the terminals in order for the message to be transmitted, and to continue the process.

The messages are transmitted between neurons via chemical messengers called neurotransmitters.[113] Neurotransmitters that are synthesized by the neuron are stored in tiny sacs called vesicles.[114] When the right message is received and processed by the neuron, the vesicles will release neurotransmitters. The neurotransmitter is released, it travels to the next neuron across a gap called the synaptic gap, and it attaches to the dendrite (i.e. the receptor). If the neurotransmitter attaches to the "right" dendrite, the dendrite will forward the appropriate message to the cell body. These transmissions work like a lock and key system, and only the correct neurotransmitters (i.e. chemicals) will "open the lock" at the dendrites to affect the second neuron.[115] The neurotransmitters that do not "open the lock" are ignored. This permits the brain to selectively receive and process the correct messages, and to not process extraneous messages. Additionally, within the neuron that releases the neurotransmitter, there are chemicals called "transporters" that regulate the signals between neurons.[116] Once the signal is successfully sent, the transporters are able to stop the signals between neurons by regulating the releases of neurotransmitters.[117]

The messages create memories.[118] Tiny memory bumps called dendritic spines grow when nerve cells are stimulated by sensory input.[119] It takes about 1,000 dendritic spines to create a single memory.[120] Memories that are linked together, and memories that are used more often, are longer-lasting and have a greater effect upon the person's thoughts. Further, emotional and stressful memories are stronger because they produce more and larger dendritic spines.[121] Therefore a strong emotional experience that is stored in the limbic system and tied to "reward" and "survival" is a stronger memory than a memory that is stored in the neo-cortex and tied to a calm and reasoned experience. What we learn and remember affects our future actions, therefore drugs that produce strong euphoria and store the memory in the limbic system will more forcefully drive thoughts than memories stored in the neo-cortex.[122] In fact, memories stored in the limbic system will be processed 4 to 5 times more quickly than memories stored in the neo-cortex.[123] Accordingly, the limbic system will drive thoughts and actions <u>before</u> the neo-cortex has a chance to produce a thought resulting from the same experience.

The release of the neurotransmitters through drug use is effective because its affects the way that neurons communicate with each other. The drugs produce chemicals that mimic the normal neurotransmitters so that they are perceived by the dendrites to be legitimate messages. The dendrites receive the messages, the messages are processed, and <u>*the brain is changed*</u> based upon the false messages produced by the drugs. The biology of the brain—which teaches us to walk and talk—is used by the addict's brain to associate drug use with pleasure and survival, and to produce thoughts that encourage and reward further drug use. The drugs "hijack" the brain. The drugs change the brain—which we use to guide our everyday lives—to reward and perpetuate drug use.

Drugs release many times the amount of dopamine that the brain can create naturally, drugs teach the brain to produce thoughts that reward drug use, and drugs change the brain. The brain adjusts to the overwhelming surges in dopamine by producing less dopamine and by reducing the number of receptors that can receive signals. As a result, over time, the addict needs more drugs to achieve the same euphoria or "high". The brain is changed and conditioned to seek large amounts of dopamine from drugs, and it drives the addict to use more drugs. In order to feel the euphoria produced by the overwhelming surges of dopamine produced by drugs, the brain produces thoughts that drive the addict to take larger amounts of drugs and to seek greater surges of dopamine. Further, the overwhelming surges in dopamine produced by drugs can cause the brain to attribute less reward and pleasure to the natural dopamine levels produced by normal pleasures and experiences (such as a good meal, love and affection, a smile, or companionship). As a result, the addict feels less pleasure from normal activities because the brain attributes less reward and less pleasure to the natural levels of dopamine produced by normal activities.

As you know from everyday experience, people think and act differently. There is no guarantee that each person will make the best decision. People have different brains, different experiences, and different lessons imprinted on their brain. The brains of different people will produce thoughts and make decisions based upon the strongest experiences and lessons that are imprinted on <u>their</u> brain, and they will use the thoughts that come first to mind within <u>their</u> brains. The addict is no different. The addict's brain has been affected by drugs, and the drugs created strong emotional memories within the brain. As a result, addict is guided by their brain to act upon the stronger and more instinctual memories that are stored in the brain—and the resulting thoughts that encourage and reward continued drug use.

This basic process allows us to process experiences and to learn. This process allows the brain to produce thoughts that are alignment with our experiences, and it allows the brain to coordinate and regulate the body's thoughts and actions.

Drugs produce extreme pleasure through the release of chemicals within the brain, and this has a direct effect upon the brain's limbic system and its control and regulation of the brain's reward functions. The positive and negative emotions produced by the limbic system direct the individual to either repeat OR to discontinue the actions and behaviors.[124] Normally, pleasure is initiated by an experience (such as a kiss or an embrace) where axons release neurotransmitters that are received by the dendrites to produce pleasure. The extent of the pleasure will vary between individuals; however, fairly consistently, the brain learns to associate these experiences with pleasure and it creates lessons that reward these behaviors with feelings of pleasure. When drugs are used, the amount of the neurotransmitters that are produced, received and processed will be many times greater than the neurotransmitters produced by normal experience.[125] Drugs can release two to ten times the amount of chemicals that the brain can naturally create to produce pleasure; and, some drugs can last much longer than the effects produced by natural rewards.[126] The drug's over-stimulation of the brain's reward system produces euphoric effects that are much more intense than the normal pleasure and rewards. The brain's limbic system *learns* to expect to receive these very intense euphoric feelings, and the brain's limbic system is changed and *taught to seek* the chemicals and the behaviors that produce these intense euphoric feelings.[127] That is, the brain is taught to seek more drug use as it attempts to reproduce the euphoria. The ill effects of drug use are minimized by the limbic system's *new* "reward system", and the addict's "hijacked" brain produces thoughts encouraging drug use which are *perceived as critical to the person's existence* and which cause the person to obtain additional drugs in order to re-produce the *necessary* euphoria caused by the drug. Indeed, *drug use is perceived as a basic survival need by the brain*.[128] The brain learns to instinctively associate the drug use with pleasure and as a critical part of everyday life, and the brain is taught to create lessons and thoughts that reward drug use. In short, not only do the drugs produce and receive more neurotransmitters, *the brain changes and learns to produce thoughts that directly link drug use with pleasure and survival*.

ENDNOTES

1   *Drugs, Brains and Behavior—The Science of Addiction.* © copyright 2010 National Institute on Drug Abuse (NIDA), p. 15. Print.

2   Inaba, Darryl, and William Cohen. *Uppers, Downers, All Arounders.* 7th Edition. Medford, Oregon: CNS Publications, Inc., 2010. 2.11-2.13. Print.

3   Inaba, Darryl, and William Cohen. *Uppers, Downers, All Arounders.* 7th Edition. Medford, Oregon: CNS Publications, Inc., 2010. 2.12-2.13. Print.

4   Inaba, Darryl, and William Cohen. *Uppers, Downers, All Arounders.* 7th Edition. Medford, Oregon: CNS Publications, Inc., 2010. 2.13. Print.

5   Inaba, Darryl, and William Cohen. *Uppers, Downers, All Arounders.* 7th Edition. Medford, Oregon: CNS Publications, Inc., 2010. 2.18-2.25. Print.

6   Inaba, Darryl, and William Cohen. *Uppers, Downers, All Arounders.* 7th Edition. Medford, Oregon: CNS Publications, Inc., 2010. 2.11-2.19. Print.

7   Inaba, Darryl, and William Cohen. *Uppers, Downers, All Arounders.* 7th Edition. Medford, Oregon: CNS Publications, Inc., 2010. 2.11-2.19. Print.

8   Inaba, Darryl, and William Cohen. *Uppers, Downers, All Arounders.* 7th Edition. Medford, Oregon: CNS Publications, Inc., 2010. 2.11-2.13. Print.

9   Inaba, Darryl, and William Cohen. *Uppers, Downers, All Arounders.* 7th Edition. Medford, Oregon: CNS Publications, Inc., 2010. 2.11-2.13. Print.

10 Inaba, Darryl, and William Cohen. *Uppers, Downers, All Arounders.* 7th Edition. Medford, Oregon: CNS Publications, Inc., 2010. 2.20-2.21 and 7.31-7.32. Print.

11 Inaba, Darryl, and William Cohen. *Uppers, Downers, All Arounders.* 7th Edition. Medford, Oregon: CNS Publications, Inc., 2010. 2.20-2.21. Print.

12 Inaba, Darryl, and William Cohen. *Uppers, Downers, All Arounders.* 7th Edition. Medford, Oregon: CNS Publications, Inc., 2010. 2.23-2.24. Print.

[13] Inaba, Darryl, and William Cohen. *Uppers, Downers, All Arounders.* 7th Edition. Medford, Oregon: CNS Publications, Inc., 2010. 3.6. Print.

[14] Inaba, Darryl, and William Cohen. *Uppers, Downers, All Arounders.* 7th Edition. Medford, Oregon: CNS Publications, Inc., 2010. 2.25-2.27. Print.

[15] Inaba, Darryl, and William Cohen. *Uppers, Downers, All Arounders.* 7th Edition. Medford, Oregon: CNS Publications, Inc., 2010. 2.22-2.27. Print.

[16] Inaba, Darryl, and William Cohen. *Uppers, Downers, All Arounders.* 7th Edition. Medford, Oregon: CNS Publications, Inc., 2010. 2.10. Print.

[17] Inaba, Darryl, and William Cohen. *Uppers, Downers, All Arounders.* 7th Edition. Medford, Oregon: CNS Publications, Inc., 2010. 2.10. Print.

[18] Inaba, Darryl, and William Cohen. *Uppers, Downers, All Arounders.* 7th Edition. Medford, Oregon: CNS Publications, Inc., 2010. 2.10. Print.

[19] Inaba, Darryl, and William Cohen. *Uppers, Downers, All Arounders.* 7th Edition. Medford, Oregon: CNS Publications, Inc., 2010. 2.10. Print.

[20] Inaba, Darryl, and William Cohen. *Uppers, Downers, All Arounders.* 7th Edition. Medford, Oregon: CNS Publications, Inc., 2010. 2.14-2.17. Print.

[21] Inaba, Darryl, and William Cohen. *Uppers, Downers, All Arounders.* 7th Edition. Medford, Oregon: CNS Publications, Inc., 2010. 2.10-2.11. Print.

[22] Inaba, Darryl, and William Cohen. *Uppers, Downers, All Arounders.* 7th Edition. Medford, Oregon: CNS Publications, Inc., 2010. 2.11-2.15. Print.

[23] Inaba, Darryl, and William Cohen. *Uppers, Downers, All Arounders.* 7th Edition. Medford, Oregon: CNS Publications, Inc., 2010. 2.11. Print.

[24] Inaba, Darryl, and William Cohen. *Uppers, Downers, All Arounders.* 7th Edition. Medford, Oregon: CNS Publications, Inc., 2010. 2.11-2.15. Print.

[25] Inaba, Darryl, and William Cohen. *Uppers, Downers, All Arounders.* 7th Edition. Medford, Oregon: CNS Publications, Inc., 2010. 2.11-2.17. Print.

[26] Inaba, Darryl, and William Cohen. *Uppers, Downers, All Arounders.* 7th Edition. Medford, Oregon: CNS Publications, Inc., 2010. 2.14. Print.

[27] Inaba, Darryl, and William Cohen. *Uppers, Downers, All Arounders.* 7th Edition. Medford, Oregon: CNS Publications, Inc., 2010. 2.11-2.17. Print.

[28] Inaba, Darryl, and William Cohen. *Uppers, Downers, All Arounders.* 7th Edition. Medford, Oregon: CNS Publications, Inc., 2010. 2.27-2.29. Print.

[29] Inaba, Darryl, and William Cohen. *Uppers, Downers, All Arounders.* 7th Edition. Medford, Oregon: CNS Publications, Inc., 2010. 2.27-2.29. Print.

[30] Inaba, Darryl, and William Cohen. *Uppers, Downers, All Arounders.* 7th Edition. Medford, Oregon: CNS Publications, Inc., 2010. 9.21-9.25. Print.

[31] Inaba, Darryl, and William Cohen. *Uppers, Downers, All Arounders.* 7th Edition. Medford, Oregon: CNS Publications, Inc., 2010. 8.8 and 9.9. Print.

32   Inaba, Darryl, and William Cohen. *Uppers, Downers, All Arounders*. 7th Edition. Medford, Oregon: CNS Publications, Inc., 2010. 2.35-2.36. Print.

33   Inaba, Darryl, and William Cohen. *Uppers, Downers, All Arounders*. 7th Edition. Medford, Oregon: CNS Publications, Inc., 2010. 2.36-2.38. Print.

34   Inaba, Darryl, and William Cohen. *Uppers, Downers, All Arounders*. 7th Edition. Medford, Oregon: CNS Publications, Inc., 2010. 2.37. Print.

35   Inaba, Darryl, and William Cohen. *Uppers, Downers, All Arounders*. 7th Edition. Medford, Oregon: CNS Publications, Inc., 2010. 2.37. Print.

36   Inaba, Darryl, and William Cohen. *Uppers, Downers, All Arounders*. 7th Edition. Medford, Oregon: CNS Publications, Inc., 2010. 2.37. Print.

37   Volpicelli, Joseph, MD and PHD, and Szalavitz, Maia. *Recovery Options*. 1st Edition. New York: John Wiley & Sons, Inc., 2000. 40-42. Print.

38   Volpicelli, Joseph, MD and PHD, and Szalavitz, Maia. *Recovery Options*. 1st Edition. New York: John Wiley & Sons, Inc., 2000. 40-42. Print.

39   Volpicelli, Joseph, MD and PHD, and Szalavitz, Maia. *Recovery Options*. 1st Edition. New York: John Wiley & Sons, Inc., 2000. 41. Print.

40   Volpicelli, Joseph, MD and PHD, and Szalavitz, Maia. *Recovery Options*. 1st Edition. New York: John Wiley & Sons, Inc., 2000. 41. Print.

41   Volpicelli, Joseph, MD and PHD, and Szalavitz, Maia. *Recovery Options*. 1st Edition. New York: John Wiley & Sons, Inc., 2000. 79-85. Print.

42   Inaba, Darryl, and William Cohen. *Uppers, Downers, All Arounders*. 7th Edition. Medford, Oregon: CNS Publications, Inc., 2010. 2.28-2.30 and 4.16-4.18. Print.

43   Inaba, Darryl, and William Cohen. *Uppers, Downers, All Arounders*. 7th Edition. Medford, Oregon: CNS Publications, Inc., 2010. 2.28-2.30 and 4.16-4.18. Print.

44   Inaba, Darryl, and William Cohen. *Uppers, Downers, All Arounders*. 7th Edition. Medford, Oregon: CNS Publications, Inc., 2010. 2.28. Print.

45   Inaba, Darryl, and William Cohen. *Uppers, Downers, All Arounders*. 7th Edition. Medford, Oregon: CNS Publications, Inc., 2010. 2.28. Print.

46   Inaba, Darryl, and William Cohen. *Uppers, Downers, All Arounders*. 7th Edition. Medford, Oregon: CNS Publications, Inc., 2010. 2.28. Print.

47   Inaba, Darryl, and William Cohen. *Uppers, Downers, All Arounders*. 7th Edition. Medford, Oregon: CNS Publications, Inc., 2010. 4.17. Print. dand Volpicelli, Joseph, MD and PHD, and Szalavitz, Maia. *Recovery Options*. 1st Edition. New York: John Wiley & Sons, Inc., 2000. 115. Print.

48   Inaba, Darryl, and William Cohen. *Uppers, Downers, All Arounders*. 7th Edition. Medford, Oregon: CNS Publications, Inc., 2010. 5.23-5.24.

Print. and Volpicelli, Joseph, MD and PHD, and Szalavitz, Maia. *Recovery Options*. 1st Edition. New York: John Wiley & Sons, Inc., 2000. 116. Print.

49 Inaba, Darryl, and William Cohen. *Uppers, Downers, All Arounders*. 7th Edition. Medford, Oregon: CNS Publications, Inc., 2010. 5.23-5.24. Print. and Volpicelli, Joseph, MD and PHD, and Szalavitz, Maia. *Recovery Options*. 1st Edition. New York: John Wiley & Sons, Inc., 2000. 116. Print.

50 Inaba, Darryl, and William Cohen. *Uppers, Downers, All Arounders*. 7th Edition. Medford, Oregon: CNS Publications, Inc., 2010. 2.28-2.30 and 4.16-4.18. Print.

51 Inaba, Darryl, and William Cohen. *Uppers, Downers, All Arounders*. 7th Edition. Medford, Oregon: CNS Publications, Inc., 2010. 2.28-2.30 and 4.16-4.18. Print.

52 Inaba, Darryl, and William Cohen. *Uppers, Downers, All Arounders*. 7th Edition. Medford, Oregon: CNS Publications, Inc., 2010. 2.28-2.30 and 4.29. Print.

53 Inaba, Darryl, and William Cohen. *Uppers, Downers, All Arounders*. 7th Edition. Medford, Oregon: CNS Publications, Inc., 2010. 4.17. Print.

54 Inaba, Darryl, and William Cohen. *Uppers, Downers, All Arounders*. 7th Edition. Medford, Oregon: CNS Publications, Inc., 2010. 4.17. Print.

55 Inaba, Darryl, and William Cohen. *Uppers, Downers, All Arounders*. 7th Edition. Medford, Oregon: CNS Publications, Inc., 2010. 4.42. Print.

56 Inaba, Darryl, and William Cohen. *Uppers, Downers, All Arounders*. 7th Edition. Medford, Oregon: CNS Publications, Inc., 2010. 8.13. Print.

57 What's Wrong with the Teenage Brain? © copyright 2012 Wall Street Journal, Print January 28, 2012.

58 What's Wrong with the Teenage Brain? © copyright 2012 Wall Street Journal, Print January 28, 2012. and Inaba, Darryl, and William Cohen. *Uppers, Downers, All Arounders*. 7th Edition. Medford, Oregon: CNS Publications, Inc., 2010. 2.32. Print.

59 Inaba, Darryl, and William Cohen. *Uppers, Downers, All Arounders*. 7th Edition. Medford, Oregon: CNS Publications, Inc., 2010. 8.12-8.13. Print.

60 Inaba, Darryl, and William Cohen. *Uppers, Downers, All Arounders*. 7th Edition. Medford, Oregon: CNS Publications, Inc., 2010. 8.12-8.13. Print.

61 Inaba, Darryl, and William Cohen. *Uppers, Downers, All Arounders*. 7th Edition. Medford, Oregon: CNS Publications, Inc., 2010. 10.1-10.33. Print.

62 Inaba, Darryl, and William Cohen. *Uppers, Downers, All Arounders*. 7th Edition. Medford, Oregon: CNS Publications, Inc., 2010. 10.1-10.33. Print.

63 Inaba, Darryl, and William Cohen. *Uppers, Downers, All Arounders*. 7th Edition. Medford, Oregon: CNS Publications, Inc., 2010. 8.12-8.19 and 8.28-8.33. Print.

64 Inaba, Darryl, and William Cohen. *Uppers, Downers, All Arounders*. 7th Edition. Medford, Oregon: CNS Publications, Inc., 2010. 8.28-8.33 and 9.32-33. Print. and Volpicelli, Joseph, MD and PHD, and Szalavitz, Maia. *Recovery Options*. 1st Edition. New York: John Wiley & Sons, Inc., 2000. 74-77. Print.

65 Inaba, Darryl, and William Cohen. *Uppers, Downers, All Arounders*. 7th Edition. Medford, Oregon: CNS Publications, Inc., 2010. 8.28-8.33 and 9.32-9.33. Print. and Volpicelli, Joseph, MD and PHD, and Szalavitz, Maia. *Recovery Options*. 1st Edition. New York: John Wiley & Sons, Inc., 2000. 74-77. Print.

66 Inaba, Darryl, and William Cohen. *Uppers, Downers, All Arounders*. 7th Edition. Medford, Oregon: CNS Publications, Inc., 2010. 9.32. Print.

67 Volpicelli, Joseph, MD and PHD, and Szalavitz, Maia. *Recovery Options*. 1st Edition. New York: John Wiley & Sons, Inc., 2000. 60-74. Print. See also Inaba, Darryl, and William Cohen. *Uppers, Downers, All Arounders*. 7th Edition. Medford, Oregon: CNS Publications, Inc., 2010. 9.19-9.25.Print.

68 Volpicelli, Joseph, MD and PHD, and Szalavitz, Maia. *Recovery Options*. 1st Edition. New York: John Wiley & Sons, Inc., 2000. 61. Print.

69 Volpicelli, Joseph, MD and PHD, and Szalavitz, Maia. *Recovery Options*. 1st Edition. New York: John Wiley & Sons, Inc., 2000. 62-68. Print.

70 Volpicelli, Joseph, MD and PHD, and Szalavitz, Maia. *Recovery Options*. 1st Edition. New York: John Wiley & Sons, Inc., 2000. 14-18. Print. and *Narcotics Anonymous*, 5th edition © copyright 1988 Narcotics Anonymous World Services, Inc.

71 Volpicelli, Joseph, MD and PHD, and Szalavitz, Maia. *Recovery Options*. 1st Edition. New York: John Wiley & Sons, Inc., 2000. 68-70. Print.

72 Volpicelli, Joseph, MD and PHD, and Szalavitz, Maia. *Recovery Options*. 1st Edition. New York: John Wiley & Sons, Inc., 2000. 70-71. Print.

73 Volpicelli, Joseph, MD and PHD, and Szalavitz, Maia. *Recovery Options*. 1st Edition. New York: John Wiley & Sons, Inc., 2000. 71-72. Print.

74 Volpicelli, Joseph, MD and PHD, and Szalavitz, Maia. *Recovery Options*. 1st Edition. New York: John Wiley & Sons, Inc., 2000. 71-72. Print.

75 Volpicelli, Joseph, MD and PHD, and Szalavitz, Maia. *Recovery Options*. 1st Edition. New York: John Wiley & Sons, Inc., 2000. 72-73. Print.

76 *Alcoholics Anonymous*, © copyright 2001 A.A. World Services, Inc.

[77] Volpicelli, Joseph, MD and PHD, and Szalavitz, Maia. *Recovery Options.* 1st Edition. New York: John Wiley & Sons, Inc., 2000. 73-74. Print.

[78] *The Next Step Toward A Better Life.* © copyright 2010 U.S. Department of Health and Human Services, Substance Abuse and Mental Health Administration (SMAHSA), p. 4. Print.

[79] Volpicelli, Joseph, MD and PHD, and Szalavitz, Maia. *Recovery Options.* 1st Edition. New York: John Wiley & Sons, Inc., 2000. 73-74. Print.

[80] Volpicelli, Joseph, MD and PHD, and Szalavitz, Maia. *Recovery Options.* 1st Edition. New York: John Wiley & Sons, Inc., 2000. 72. Print.

[81] Goode, Erich. *Drugs in American Society.* 4th Edition. New York: McGraw-Hill, 2012. 60-83. Print. and Volpicelli, Joseph, MD and PHD, and Szalavitz, Maia. *Recovery Options.* 1st Edition. New York: John Wiley & Sons, Inc., 2000. 50-51. Print.

[82] Dick, Danielle, PHD, and Agrawal, Arpana. *The Genetics of Alcohol and Other Drug Dependence.* © copyright 2008 National Institute on Drug Abuse (NIDA), p. 111.

[83] Inaba, Darryl, and William Cohen. *Uppers, Downers, All Arounders.* 7th Edition. Medford, Oregon: CNS Publications, Inc., 2010. 10.4. Print.

[84] Inaba, Darryl, and William Cohen. *Uppers, Downers, All Arounders.* 7th Edition. Medford, Oregon: CNS Publications, Inc., 2010. 2.30 and 10.8-10.9. Print.

[85] Inaba, Darryl, and William Cohen. *Uppers, Downers, All Arounders.* 7th Edition. Medford, Oregon: CNS Publications, Inc., 2010. 10.8-10.13. Print.

[86] Inaba, Darryl, and William Cohen. *Uppers, Downers, All Arounders.* 7th Edition. Medford, Oregon: CNS Publications, Inc., 2010. 8.28-8.30 and 4.16-4.18. Print.

[87] Inaba, Darryl, and William Cohen. *Uppers, Downers, All Arounders.* 7th Edition. Medford, Oregon: CNS Publications, Inc., 2010.8.28-8.30 and 4.16-4.18. Print.

[88] Inaba, Darryl, and William Cohen. *Uppers, Downers, All Arounders.* 7th Edition. Medford, Oregon: CNS Publications, Inc., 2010. 8.28-8.30 and 4.16-4.18. Print.

[89] Goode, Erich. *Drugs in American Society.* 4th Edition. New York: McGraw-Hill, 2012. 69-78. Print.

[90] Goode, Erich. *Drugs in American Society.* 4th Edition. New York: McGraw-Hill, 2012. 60-83. Print. and Volpicelli, Joseph, MD and PHD, and Szalavitz, Maia. *Recovery Options.* 1st Edition. New York: John Wiley & Sons, Inc., 2000. 50-51. Print.

91 Inaba, Darryl, and William Cohen. *Uppers, Downers, All Arounders.* 7th Edition. Medford, Oregon: CNS Publications, Inc., 2010. 5.22-5.24 and 5.43. Print.
92 Volpicelli, Joseph, MD and PHD, and Szalavitz, Maia. *Recovery Options.* 1st Edition. New York: John Wiley & Sons, Inc., 2000. 72-73. Print.
93 Volpicelli, Joseph, MD and PHD, and Szalavitz, Maia. *Recovery Options.* 1st Edition. New York: John Wiley & Sons, Inc., 2000. 134-142. Print.
94 Volpicelli, Joseph, MD and PHD, and Szalavitz, Maia. *Recovery Options.* 1st Edition. New York: John Wiley & Sons, Inc., 2000. 135-136. Print.
95 Volpicelli, Joseph, MD and PHD, and Szalavitz, Maia. *Recovery Options.* 1st Edition. New York: John Wiley & Sons, Inc., 2000. 138. Print.
96 Volpicelli, Joseph, MD and PHD, and Szalavitz, Maia. *Recovery Options.* 1st Edition. New York: John Wiley & Sons, Inc., 2000. 134-142. Print.
97 Volpicelli, Joseph, MD and PHD, and Szalavitz, Maia. *Recovery Options.* 1st Edition. New York: John Wiley & Sons, Inc., 2000. 146-148. Print.
98 Volpicelli, Joseph, MD and PHD, and Szalavitz, Maia. *Recovery Options.* 1st Edition. New York: John Wiley & Sons, Inc., 2000. 149-151. Print.
99 Volpicelli, Joseph, MD and PHD, and Szalavitz, Maia. *Recovery Options.* 1st Edition. New York: John Wiley & Sons, Inc., 2000. 151-154. Print.
100 Volpicelli, Joseph, MD and PHD, and Szalavitz, Maia. *Recovery Options.* 1st Edition. New York: John Wiley & Sons, Inc., 2000. 91, 127-129 and 152. Print. Inaba, Darryl, and William Cohen. *Uppers, Downers, All Arounders.* 7th Edition. Medford, Oregon: CNS Publications, Inc., 2010. 4.26-4.29.
101 Volpicelli, Joseph, MD and PHD, and Szalavitz, Maia. *Recovery Options.* 1st Edition. New York: John Wiley & Sons, Inc., 2000. 127-129. Print. and Inaba, Darryl, and William Cohen. *Uppers, Downers, All Arounders.* 7th Edition. Medford, Oregon: CNS Publications, Inc., 2010. 4.26-4.29.
102 Volpicelli, Joseph, MD and PHD, and Szalavitz, Maia. *Recovery Options.* 1st Edition. New York: John Wiley & Sons, Inc., 2000. 127-129. Print. and Inaba, Darryl, and William Cohen. *Uppers, Downers, All Arounders.* 7th Edition. Medford, Oregon: CNS Publications, Inc., 2010. 4.26-4.29.
103 Volpicelli, Joseph, MD and PHD, and Szalavitz, Maia. *Recovery Options.* 1st Edition. New York: John Wiley & Sons, Inc., 2000. 127-129. Print. and Inaba, Darryl, and William Cohen. *Uppers, Downers, All Arounders.* 7th Edition. Medford, Oregon: CNS Publications, Inc., 2010. 4.26-4.29.
104 Volpicelli, Joseph, MD and PHD, and Szalavitz, Maia. *Recovery Options.* 1st Edition. New York: John Wiley & Sons, Inc., 2000. 127-129. Print.

and Inaba, Darryl, and William Cohen. *Uppers, Downers, All Arounders.* 7th Edition. Medford, Oregon: CNS Publications, Inc., 2010. 4.26-4.29.

[105] Inaba, Darryl, and William Cohen. *Uppers, Downers, All Arounders.* 7th Edition. Medford, Oregon: CNS Publications, Inc., 2010. 4.29. Print.

[106] Volpicelli, Joseph, MD and PHD, and Szalavitz, Maia. *Recovery Options.* 1st Edition. New York: John Wiley & Sons, Inc., 2000. 132. Print. and Inaba, Darryl, and William Cohen. *Uppers, Downers, All Arounders.* 7th Edition. Medford, Oregon: CNS Publications, Inc., 2010. 9.21.

[107] Volpicelli, Joseph, MD and PHD, and Szalavitz, Maia. *Recovery Options.* 1st Edition. New York: John Wiley & Sons, Inc., 2000. 176. Print

[108] *Three Views of Al-Anon—Alcoholics Speak to the Family.* © copyright 2010 Al-Anon, Pamphlet. Print.

[109] Volpicelli, Joseph, MD and PHD, and Szalavitz, Maia. *Recovery Options.* 1st Edition. New York: John Wiley & Sons, Inc., 2000. 48. Print

[110] Inaba, Darryl, and William Cohen. *Uppers, Downers, All Arounders.* 7th Edition. Medford, Oregon: CNS Publications, Inc., 2010. 2.1-2.49. Print.

[111] Inaba, Darryl, and William Cohen. *Uppers, Downers, All Arounders.* 7th Edition. Medford, Oregon: CNS Publications, Inc., 2010. 2.19. Print.

[112] Inaba, Darryl, and William Cohen. *Uppers, Downers, All Arounders.* 7th Edition. Medford, Oregon: CNS Publications, Inc., 2010. 2.18-2.25. Print.

[113] Inaba, Darryl, and William Cohen. *Uppers, Downers, All Arounders.* 7th Edition. Medford, Oregon: CNS Publications, Inc., 2010. 2.19. Print.

[114] Inaba, Darryl, and William Cohen. *Uppers, Downers, All Arounders.* 7th Edition. Medford, Oregon: CNS Publications, Inc., 2010. 2.19. Print.

[115] Inaba, Darryl, and William Cohen. *Uppers, Downers, All Arounders.* 7th Edition. Medford, Oregon: CNS Publications, Inc., 2010. 2.18-2.20. Print.

[116] Inaba, Darryl, and William Cohen. *Uppers, Downers, All Arounders.* 7th Edition. Medford, Oregon: CNS Publications, Inc., 2010. 2.19-2.25. Print.

[117] Inaba, Darryl, and William Cohen. *Uppers, Downers, All Arounders.* 7th Edition. Medford, Oregon: CNS Publications, Inc., 2010. 2.18-2.23. Print.

[118] Inaba, Darryl, and William Cohen. *Uppers, Downers, All Arounders.* 7th Edition. Medford, Oregon: CNS Publications, Inc., 2010. 2.11-2.14. Print.

[119] Inaba, Darryl, and William Cohen. *Uppers, Downers, All Arounders.* 7th Edition. Medford, Oregon: CNS Publications, Inc., 2010. 2.12-2.13. Print.

[120] Inaba, Darryl, and William Cohen. *Uppers, Downers, All Arounders.* 7th Edition. Medford, Oregon: CNS Publications, Inc., 2010. 2.13. Print.

121 Inaba, Darryl, and William Cohen. *Uppers, Downers, All Arounders.* 7th Edition. Medford, Oregon: CNS Publications, Inc., 2010. 2.12-2.14. Print.

122 Inaba, Darryl, and William Cohen. *Uppers, Downers, All Arounders.* 7th Edition. Medford, Oregon: CNS Publications, Inc., 2010. 2.10-2.18. Print.

123 Inaba, Darryl, and William Cohen. *Uppers, Downers, All Arounders.* 7th Edition. Medford, Oregon: CNS Publications, Inc., 2010. 2.11. Print.

124 Inaba, Darryl, and William Cohen. *Uppers, Downers, All Arounders.* 7th Edition. Medford, Oregon: CNS Publications, Inc., 2010. 2.14-2.18. Print.

125 Inaba, Darryl, and William Cohen. *Uppers, Downers, All Arounders.* 7th Edition. Medford, Oregon: CNS Publications, Inc., 2010. 2.24, 3.6 and 4.13. Print.

126 Inaba, Darryl, and William Cohen. *Uppers, Downers, All Arounders.* 7th Edition. Medford, Oregon: CNS Publications, Inc., 2010. 3.6. Print.

127 Inaba, Darryl, and William Cohen. *Uppers, Downers, All Arounders.* 7th Edition. Medford, Oregon: CNS Publications, Inc., 2010. 2.10-2.18. Print.

128 Inaba, Darryl, and William Cohen. *Uppers, Downers, All Arounders.* 7th Edition. Medford, Oregon: CNS Publications, Inc., 2010. 2.10-2.18 Print.

ABOUT THE AUTHORS

We have graduate degrees from top schools, and we worked at Fortune 500 companies. However, we couldn't use our existing knowledge to address our son's addiction—we had to acquire new knowledge and skills, and develop our ability to apply the new knowledge and skills in real-life situations. The purpose of this book is to share this new knowledge and the skills with you.